Holographic Materials and Applications

Edited by Manoj Kumar

Published in London, United Kingdom

IntechOpen

Supporting open minds since 2005

Holographic Materials and Applications
http://dx.doi.org/10.5772/intechopen.77488
Edited by Manoj Kumar

Contributors
Manoj Kumar, Yassine Tounsi, Abdelkrim Nassim, Fernando Mendoza-Santoyo, Abdulatef Ghlaifan, Tomoya Nakamura, Shunsuke Takahashi, Shinji Kimura, Kazuhiko Takahashi, Yuji Aburakawa, Shunsuke Igarashi, Shiho Torashima, Masahiro Yamaguchi, Nam Kim, Hui-Ying Wu, Chang-Won Shin, Michal Makowski, Nikolay Nikonorov, Sergei Ivanov, Viktor Dubrovin, Dmitii Kliukin, Luis Luevano, Eduardo Lopez, Hector Quintero, Gokhan Bora Esmer, William Alschuler

First published in London, United Kingdom, 2019 by IntechOpen
IntechOpen is the global imprint of INTECHOPEN LIMITED, registered in England and Wales, registration number: 11086078, The Shard, 25th floor, 32 London Bridge Street
London, SE19SG - United Kingdom
Printed in Croatia

British Library Cataloguing-in-Publication Data
A catalogue record for this book is available from the British Library

Additional hard and PDF copies can be obtained from orders@intechopen.com

Holographic Materials and Applications
Edited by Manoj Kumar
p. cm.
Print ISBN 978-1-78984-788-8
Online ISBN 978-1-78984-789-5
eBook (PDF) ISBN 978-1-83962-223-6

Meet the editor

Manoj Kumar obtained a BSc (Hons.) and MSc in Physics from the University of Delhi in 2006 and 2008, respectively; an M.Tech degree in Instrumentation from the National Institute of Technology, Kurukshetra, in 2011, and a PhD in Optical Engineering from the Indian Institute of Technology Delhi, New Delhi, in 2016. Until 2017, he served as a post-doctoral research fellow in the Department of Electrical and Computer Engineering at Ben-Gurion University of the Negev, Beersheva, Israel. Presently, he is a post-doctoral research fellow at Kobe University, Japan. His research interests include 3-D imaging, digital holography, speckle metrology, and optical instrumentation. He is a member of the Optical Society of America (OSA).

Contents

Preface

Optical techniques have enormous capabilities and potentials, and the field continues to experience rapid growth. The inventions of the laser and holography diversified optical techniques, making them more powerful and easier to use in a wide range of areas, including engineering, scientific, and industrial applications. Holography added new dimensions to optical techniques and systems by offering the capability to preserve the three-dimensional (3D) details of an object. It is a powerful tool with a wide domain of applications. There has been an increased interest in the developments surrounding holographic materials and systems in order to make the field more authoritative and efficient for a large number of applications.

This volume reflects some basic concepts of holographic materials and systems while providing a comprehensive introduction to the field. Chapters include information on volume holograms, display holograms, full-color holographic optical elements, computer-generated holograms, digital imaging devices, and image processing algorithms to improve the reconstruction process and the quality of retrieved images.

This volume is intended for researchers in the field and other interested readers.

Dr. Manoj Kumar
Department of Systems Science,
Graduate School of System Informatics,
Kobe University,
Japan.

Section 1

Holographic Materials

Chapter 1

Features of Volume Holograms in Fluoride and Chloride Photothermorefractive Glass

Nikonorov Nikolay, Ivanov Sergei, Dubrovin Victor and Klyukin Dmitry

Abstract

Today, silicate photothermorefractive (PTR) glasses are well known as a holographic medium for fabrication of holographic volume diffractive optical elements. The photothermoinduced crystallization process is used for recording high-efficiency phase volume holograms in this material. These holograms are used for developing unique diffractive optical elements that provide new opportunities for the laser technique, for example, narrowband filters for solid-state lasers and laser diodes, beam combiners, holographic collimator sights, chirped gratings for laser pulse compression, etc. By now, the photothermoinduced crystallization and properties of the PTR glass are investigated well enough. However, there are some issues and features still, which are solved in the present work. The mechanism of refractive index change in fluoride photothermorefractive glass during photothermoinduced crystallization and refractive index profile of the volume Bragg gratings were discussed. We studied a fine structure of a core-shell system inside fluoride PTR glass in which a silver nanoparticle presents the core and crystalline phases of silver bromide and sodium fluoride present the shell. We report on the optical properties of volume Bragg gratings in chloride PTR glass after femtosecond laser bleaching. We demonstrated that the bleaching procedure significantly reduces the absorption and increases the thermal stability of the Bragg gratings.

Keywords: photothermorefractive glass, holography, Bragg grating, crystallization, bleaching

1. Introduction

Photothermorefractive (PTR) glasses are a new class of photosensitive materials intended for recording three-dimensional phase holograms. This glass originates from photosensitive sodium zinc aluminosilicate glasses that were applied by Stookey and Pierson in Corning, Inc. (USA) in 1977 for the first time and were referred to as polychromatic glass [1–4] worldwide and by Tsekhomsky in Vavilov State Optical Institute (Russia) [5, 6] as multichromatic (MC) glass. In the late 1980s to early 1990s [7–10], Glebov and Nikonorov in Vavilov State Optical Institute (Russia) were first proposed to implement MC glass for volume holography. This chapter is focused on the photothermoinduced crystallization process itself and the enhancement of the refractive index contrast between irradiated and nonirradiated areas. As a result, a new class of materials was developed, which is denoted [10] by a specific term such as "photothermorefractive (PTR) glass" (i.e., glass wherein after UV irradiation and subsequent thermal treatment changes in refractive index occur). Later this term started to be used worldwide [11, 12].

Nowadays, there is an increased interest in the volume Bragg gratings recorded on PTR glass due to their outstanding properties. The main advantage of these glasses is their unique combination of working characteristics such as high spectral selectivity and operating angel, great mechanical and optical strength, and high chemical durability. A great number of optical elements have been developed based on PTR glass, including extra narrow-band filters, combiners of high-intensity light beams, wavelength division multiplexing (WDM) devices, filters for increasing the spectral brightness of laser diodes, chirped gratings for compressing the light impulses, etc.

By now, the photothermoinduced crystallization and properties of the PTR glass are investigated well enough. For example, the fluoride PTR glass was designed and synthesized in ITMO University, Russia [13]. It is a photosensitive multicomponent sodium-zinc-alumina-silicate glass containing fluorine (6 mol.%) and a small amount of bromine (0.5 mol.%). Also, the PTR glass is doped with cerium, antimony, and silver. These additives are responsible for the photothermoinduced crystallization process resulting in precipitation of silver nanoparticles and sodium fluoride crystals in the volume of the glass [2, 14, 15]. Untreated fluoride PTR glass is transparent in a wide spectral range of 250–2500 nm. The photothermoinduced crystallization process in the fluoride PTR glass consists of three stages: (i) a formation of neutral silver molecular clusters after UV irradiation into the Ce^{3+} absorption band, (ii) a formation of the silver nanoparticles under the subsequent heat treatment of the UV-irradiated PTR glass near the glass transition temperature (T_g) [2], and (iii) a precipitation of silver bromide shells on the silver nanoparticles [16] and then a growth of sodium fluoride nanocrystals on them in shape of cones under the thermal treatment at temperatures above T_g [2, 17]. The basic optical and spectral properties of PTR glass are described in [2, 15, 18, 19].

In comparison with conventional holographic media (photopolymers, silver halides, gelatins, etc.), the fluoride PTR glass and the holographic optical elements (HOE)s recorded on the glass have outstanding operating characteristics. In particular, they demonstrate the high thermal stability of the HOEs and high resistance to the optical and ionizing irradiation. The optical and spectral parameters of the HOEs and gradient index (GRIN) elements do not change after its multiple heating to high enough temperature (500°C). HOEs recorded on the PTR glass show the high chemical stability and mechanical and optical resistance. From this point of view, the fluoride PTR glass reveals practically no difference with the commercial optical glass BK7 (Schott). The HOEs have a very high duration of life (dozens of years). The key advantages of fluoride PTR glass as a holographic medium are the following:

i. high optical homogeneity (the refraction index fluctuations across the glass bulk are of the order of 10^{-5}),

ii. high reproducibility of parameters of HOEs with high diffraction efficiency (above 99%),

iii. variety of technologies of glass processing—the fluoride PTR glass (like commercial optical glass) can be subjected to various kinds of mechanical grinding and polishing and another shaping technology like pressing, molding, drawing fiber, as well as laser treatment, and ion exchange,

iv. a simple synthesis of fluoride PTR glass allows us to fabricate the glass both in the laboratory conditions (hundreds of grams) and industrial ones (hundreds of kilograms). The chemical reagents required for glass fabrication are commercially available and not too expensive.

It should be noted that some features of fluoride PTR glass are unusual in comparison with other holographic media. The PTR glass contains sodium ions. These ions can be changed on another monovalent ion (silver, potassium, rubidium, and cesium) by a well-known ion-exchange technique. This approach allowed us to fabricate ion-exchanged optical, luminescent, and plasmonic waveguides as well as to increase the mechanical, thermal, and optical resistance of the glass [20].

In case of fluoride PTR glass, the photothermoinduced crystallization results in a negative change of the refractive index (RI) in the UV-irradiated area in comparison with that of the nonirradiated area ($\Delta n = -1.0 \times 10^{-3}$). We managed to control the sign of the RI increment by changing the type of halide (fluoride to chloride) in the PTR glass composition. The substitution of fluorine by chlorine resulted in the precipitation of nanocrystalline phases of mixed silver and sodium chlorides in glass host after UV irradiation and thermal treatment, which leads to a positive RI changes $\Delta n = +1.0 \times 10^{-3}$. It should be noted that positive refractive index change in chloride PTR glass provides the opportunity not only for Bragg gratings but also for the optical waveguides recording.

Untreated chloride PTR glass has the same transparency as fluoride one. And the photothermoinduced crystallization process is similar to the fluoride glass. The first stage is a generation of molecular silver clusters during UV irradiation, and the second stage is a formation of silver nanoparticles under heat treatment. On the following third stage, the heat treatment of PTR glass at temperatures above T_g leads to the precipitation of silver nanoparticles with a shell consisting of mixed sodium and silver chlorides [19, 21]. The absorption coefficient of silver nanoparticles in chloride PTR glass (at 450 nm) is very high and exceeds 100 cm^{-1}. This huge absorption is a key disadvantage chloride PTR. Despite this, the chloride PTR glass exhibits a rather low level of scattering. This is due to the small sizes of silver nanoparticles (about 3 nm) and small size of silver and sodium chloride nanocrystals (less 27 nm) [19]. This fact allowed us to conclude that the chloride PTR glass is still very attractive recording medium for photonics applications and for HOEs.

Despite the photothermoinduced crystallization and properties of the PTR glass are investigated well enough, there are some issues still, which restrict a wide application of the PTR glass. This chapter is a survey of recent achievements of ITMO University (St. Petersburg, Russia) that are focused on the investigation of the following issues: (i) mechanism of refractive index change in fluoride PTR glass during photothermoinduced crystallization, (ii) refractive index profile of the volume Bragg gratings, (iii) fine structure of core-shell system inside fluoride PTR glass in which a silver nanoparticle presents the core and crystalline phases of silver bromide and sodium fluoride present the shell, (iv) bleach of volume Bragg gratings in chloride PTR glass for decreasing optical losses and improving their thermal stability.

2. Mechanism of refractive index change in fluoride PTR glass during photothermoinduced crystallization

According to the calculations presented in [22], the residual stresses are the main reason of RIC in fluoride PTR glass. However, according to Nikonorov et al. [14], this effect is assumed to be due to the difference of refractive indexes of sodium fluoride nanocrystals and PTR glass. Recently, we conducted an experiment that proved

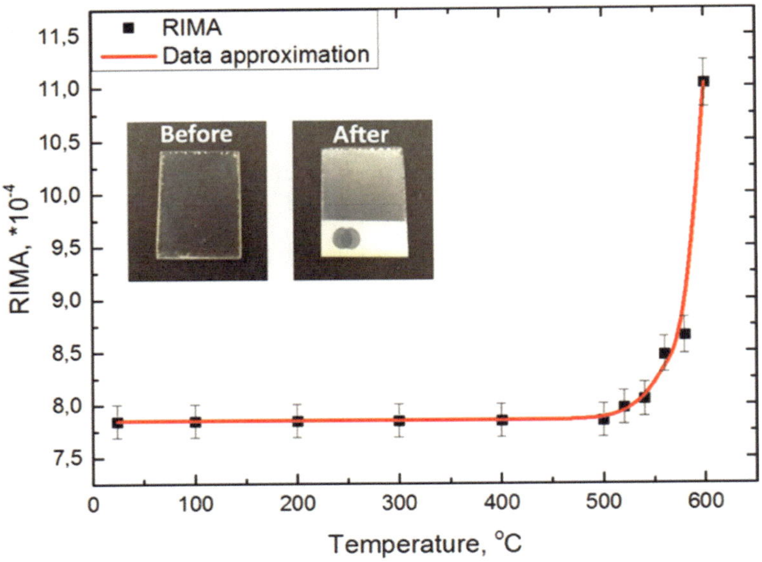

Figure 1.
Refractive index modulation amplitude evolution depending on the sample temperature (photos of the sample before and after the experiment on the insert).

that the origin of refractive index change in the PTR glass arises from the refractive index difference between PTR glass and NaF crystals.

In the experiment, transmission grating was heated in the chamber up to 600°C with 100°C step, while angular selectivity contour and diffraction efficiency (DE) were measured. Analysis of the selectivity contour revealed that heating up to the 500°C had no measurable effect on the grating DE and therefore on refractive index modulation amplitude (RIMA).

Also, it had been shown that DE of the grating starts to oscillate during the heating procedure as soon as the temperature reaches 480°C, which is due to the crystal growth process that begins around 480°C [21] and leads to RIMA value rapid increase with temperature (**Figure 1**).

Based on the results of our experiment, it is reasonable to conclude that the mechanical residual stresses cannot be the main reason of the RIC in PTR glass as even at 550°C, that is, 80°C higher than T_g, recorded Bragg grating is still functioning. Thus, the main reason for the refractive index contrast between exposed and unexposed areas is the refractive index difference between unperturbed glass and composite with NaF inclusions (**Figure 2**).

To estimate the RIC for this case, Maxwell Garnett theory [30] that allows to calculate the refractive index of the composite medium with the low volume fraction of inclusions had been applied.

To calculate the composite parameters, a model of needle NaF inclusions in the glass matrix had been chosen. In this case, the effective dielectric constant of the composite can be estimated according to the following equation [30]:

$$\delta = \left(\frac{\varepsilon - \varepsilon_h}{\varepsilon_i - \varepsilon_h}\right)\left(\frac{5\varepsilon_i + \varepsilon_h}{5\varepsilon + \varepsilon_h}\right)^{2/5} \tag{1}$$

where δ is the NaF volume fraction, ε is the effective dielectric constant of a composite, ε_i is the dielectric constant of the NaF inclusions, and ε_h is the dielectric constant of the glass matrix.

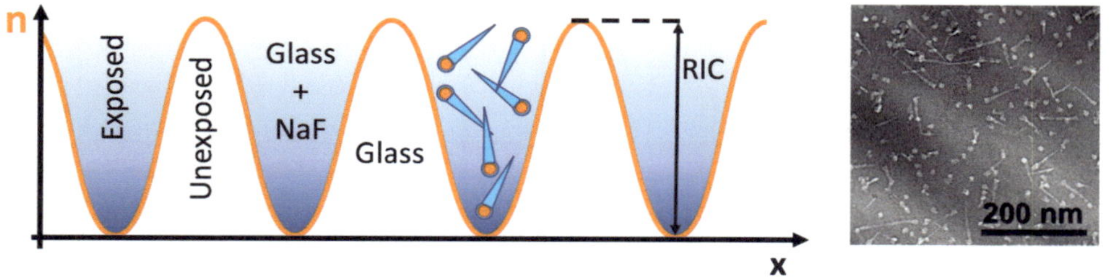

Figure 2.
Schematic representation of the volume Bragg grating (left) and TEM image of the actual grating fringe (right).

NaF volume fraction was calculated from the TEM image of the grating fringe (**Figure 2(right)**). Estimated difference of the refractive indices of such composite and unperturbed glass gave us 27.62 × 10^{-4}. It is clearly seen that this value is of the same order as the RIC in fluoride PTR glass after standard development procedure.

For experimental determination of RIC in the grating, five diffraction orders have been measured. Sum of all harmonics equals to the refractive index difference between exposed and unexposed areas and corresponds to 27.92 × 10^{-4}. At the same time, estimations based on Maxwell-Garnet theory gives us 27.62 × 10^{-4} that is close enough to the value obtained by the optical method. The difference between two methods can be attributed to rough approximations of the NaF crystal surrounding matrix refractive index as well as the fact that heat treatment leads to the changes in the refractive index of the unexposed glass [15].

Thus, it is shown that refractive index change in the exposed area appears primarily due to the NaF nanocrystal inclusions in the glass matrix, which according to the Maxwell-Garnet theory decreases the effective refractive index of such composite. It is shown experimentally that residual stresses have no measurable effect on the refractive index change.

3. Refractive index profile of the volume Bragg gratings

To record volume Bragg grating with holographic technique, one needs to interfere with two coherent beams inside the photosensitive medium. This interference will create a stationary wave with sinusoidal spatial profile of intensity. Next, a periodic variation of refractive index is inflicted by the process specific for each material. In our case, this process is used to be called photothermoinduced crystallization. Spatial distribution of the crystals results in a spatial modulation of the refractive index. In the case of nonsinusoidal gratings, refractive index profile can be expanded into a Fourier series where each harmonic represents the higher orders of diffraction. And of course, the more orders one sees, the less sinusoidal the gratings are. To utilize the whole refractive index change potential of the PTR glass, one needs to record pure sinusoidal gratings. We study the distribution of the RIC among the harmonics depending on the exposure dosage (**Figure 3(a)**).

As one can see, gratings with exposure of 0.5 J/cm^2 have a nearly pure sinusoidal profile, 96.6% of RIC is in the first harmonic (**Figure 4**). It is expected that with an increase of exposure, fraction of the RIC in the first harmonic falls. However, it does not drop below 32% even when exposure doses are enormous like 16 J/cm^2.

It is reasonable to propose that the grating profile may be affected by the scattered light; however, in such case, fraction of the RIC at the first harmonic would fall with exposure linearly. However, **Figure 3(a)** represents the saturation effect. Thus, the fall of the first harmonic fraction is not related to the scattered light during the recording process.

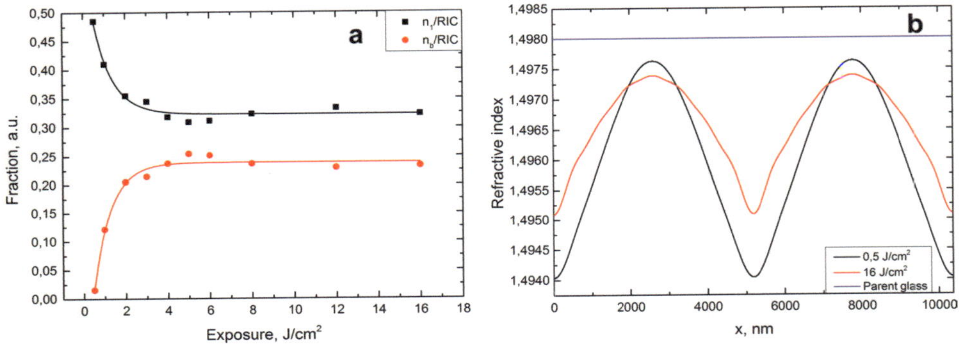

Figure 3.
(a) Dependence of the RIC distribution between harmonics on dosage; (b) comparison of refractive index profile in the holographic grating with exposure 0.5 and 16 J/cm².

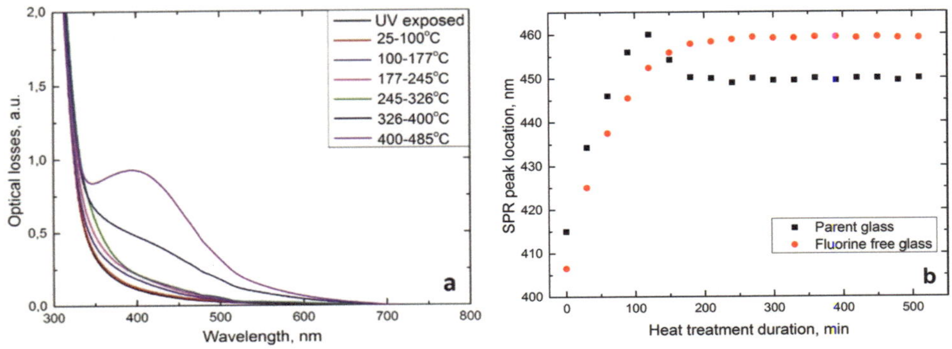

Figure 4.
(a) Absorption spectrum dynamic of the UV irradiated and heated up to 485°C for 90 min fluoride PTR glass. (b) Dynamics of SPR peak location depending on the heat treatment duration for the parent PTR glass and glass with the lowest possible concentration of fluorine.

In [23], such effect was modeled considering the saturation of the RIC on exposure. Such saturation would lead to broadening of the grating fringes and decreasing the grating contrast. This situation can occur than in the interference pattern, maximum exposure already reached its saturation and thus does not contribute to further increase RIC. In this case, further exposure leads to a decrease in the contrast between the dark and the bright regions in terms of refractive index. However, our experiments show that fringes in the gratings with big exposure are narrower than that in gratings with small exposures **Figure 3(b)**. We suppose that the nonsinusoidal profile is a complex problem and cannot be explained with the saturation of exposure.

4. Fine structure of core-shell system inside fluoride PTR glass

It is clear that (i) during photothermoinduced (PTI) crystallization process, the silver nanoparticles (NP) precipitate first (ii) that initiate sodium fluoride nanocrystals growth that is responsible for the refractive index changes. At the same time, the period between silver nanoparticles precipitation and sodium fluoride crystals growth is still unclear.

In **Figure 4(a)**, the evolution of the UV-irradiated fluoride PTR glass spectrum in the course of the heating process (heating up to 485°C for 90 min) is shown. It is seen clearly that until 300°C, no significant changes appear in the visible absorption spectrum. Some processes occurring in this temperature range manifest themselves mostly in the UV. Such behavior had been discussed a lot in [18, 24] and attributed to the cerium and antimony ions in different valency. Further heating the sample leads to the formation of silver molecular clusters, thus causing an absorption increase in the visible. After it, when a temperature close to T_g is reached, it is seen that the SPR absorption peak corresponded to silver NPs manifesting in around 400 nm is already formed.

When starting the procedure of heat treating the sample, one can observe a significant redshift of the SPR peak location (**Figure 6**). Such behavior of the SPR band can be explained by the occurrence of the shell with a high refractive index on the silver nanoparticle.

During the photothermoinduced crystallization process of silver nanoparticles, precipitation occurs first, which is followed by formation of highly refractivity phase (silver bromide) when compared with glass matrix (1.498) [21]. It should be noted that after SPR peak location reaches 460 nm, it starts moving toward the shorter wavelengths up to 450 nm, and there, it remains independently from treatment duration time increase. The blueshift indicates the precipitation of NaF crystals that reduce the refractive index of the core-shell system surrounding. Stabilization of the peak location is due to the fact that NaF crystals already act as hosts for the core-shell structures. In the case of PTR glass without fluorine subjected to the same heat treatment procedure, the only redshift of the SPR resonance peak is observed (fluorine-free glass, **Figure 6(b)**). Hence, it can be concluded that the blue shift is inflicted by a decrease in the refractive index of the core-shell surrounding.

To estimate the size of silver nanoparticles, Lorenz-Mie scattering theory [8] was used.

The observed halfwidth of the plasmon resonance band at 415 nm equals to 100 nm that corresponds to the calculated average particle size of about 2.5 nm, which leads to the resonance wavelength of 407 nm. However, the SPR peak position of the parent glass heated up to the glass transition temperature is 415 nm (**Figure 4(b)**). This should mean that at this stage, NP is already in the shell.

Our simulation with silver bromide shell shows that several initial steps in the redshift cannot be achieved with this kind of crystal. A contrast in the refractive index between the silver bromide and the glass matrix is too high. Thus, we assumed that the refractive index of the shell is less than that of AgBr crystal.

On the other hand, initial SPR peak position (415 nm) can be modeled with 2.42 nm shell thickness of pure NaBr crystal. However, refractive index contrast between the NaBr crystal and PTR glass does not allow to achieve redshift of the SPR to the 460 nm. Considering these facts, we suppose that the shell should be of mixed composition. Whereby composition of the shell can be either a solid solution of NaBr and AgBr crystals or a glass-crystal composite where AgBr inclusions present in the glass matrix. These assumptions lead to a shell with a refractive index in the range of 1.68–2.25 depending on the AgBr fraction.

TEM image of the three-layer system (**Figure 5(a)**) provides us the outer diameter of the system, which can be roughly estimated to be 10 nm. This image was obtained from a sample with 460 nm SPR peak position. Thus, we conclude that 10 nm system diameter corresponds to the case where NaF is a host for a core-shell structure. Such a condition can be met only if the NaF layer thickness is above 3.17 nm. Considering the NP diameter of 2.5 nm, we adjudge that shell thickness equals to 0.57 nm (**Figure 5(b)**). It is a special mention that shell thickness correlates to the lattice constant of AgBr (0.597 nm) and NaBr (0.578 nm) crystals [23].

Since 0.57 nm is finite shell thickness, which already corresponds to the crystal lattice size, we will assume that the shell thickness is constant during the heat treatment procedure with the growing refractive index due to an increase in the AgBr concentration either in the solid solution or in the crystal-glass mixture.

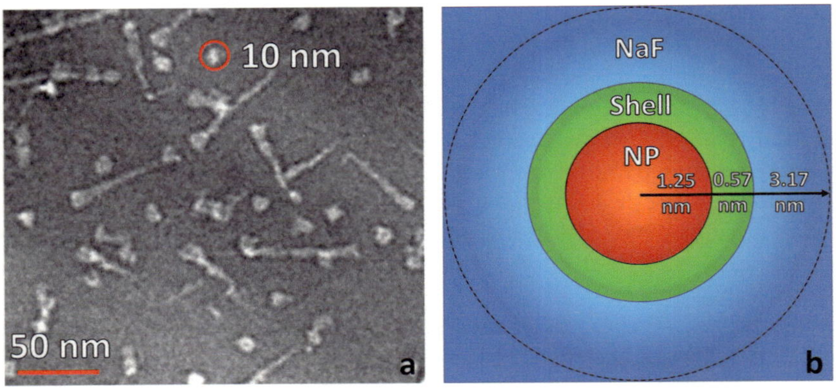

Figure 5.
(a) TEM image of a grating fringe in conventional PTR glass and (b) three-layer system model according to our estimations and the TEM image of a real system.

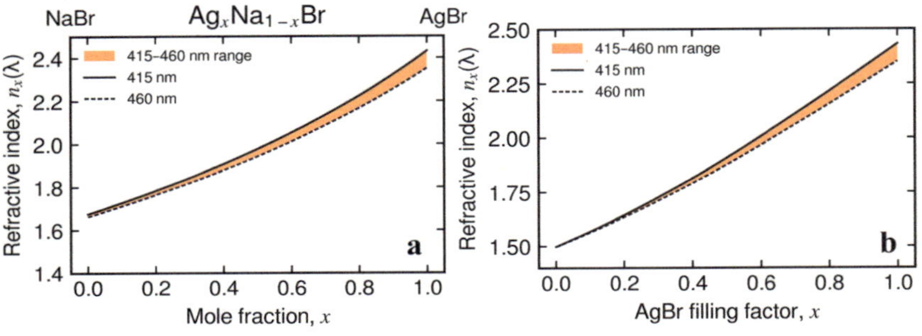

Figure 6.
Estimated refractive index of the shell for solid solution (left) and Bruggeman (right) models.

The first approach implies that the shell of our three-layer system is a solid solution of AgBr and NaBr crystal where during the heat treatment procedure, concentration of the AgBr increases. Refractive index of this system was calculated according to Varotsos [25] and its variation can be seen from **Figure 6(a)**.

It can be seen from **Figure 6(a)** that initial SPR peak position at 415 nm corresponds to 1.8% AgBr concentration in the solid solution. Thus, according to this model, initial position of the SPR equals to almost pure NaBr shell (with respect to the accuracy of the measurement and deconvolution). On the other hand, the final peak position equals to the 88.7% of the AgBr in the solid solution; therefore, at the end of this process according to this model, shell still represents a solid solution of the two crystals.

According to the second approach, a constant layer of the glass around the silver NP is filling up with AgBr crystal cells during the heat treatment. Since the refractive index of AgBr is higher than that of the surrounding glass, this process leads to the effective refractive index increase of the layer. Refractive index of this layer can be estimated with Bruggeman's two-component effective model [26]. According to our calculation, SPR peak position of 415 nm corresponds to 25% of AgBr inclusion volume fraction (**Figure 6(b)**); on the other hand, that of 460 nm corresponds to 89% of inclusion. Thus, one can see that precipitation speed of AgBr on the NP decreases with the crystal concentration growth. Accepting this model leads to conclusions that (i) the precipitation of AgBr is the most rapid at the beginning of PTI crystallization process and (ii) there is a probability that, in the course of the NP formation, the silver bromide can already be present nearby.

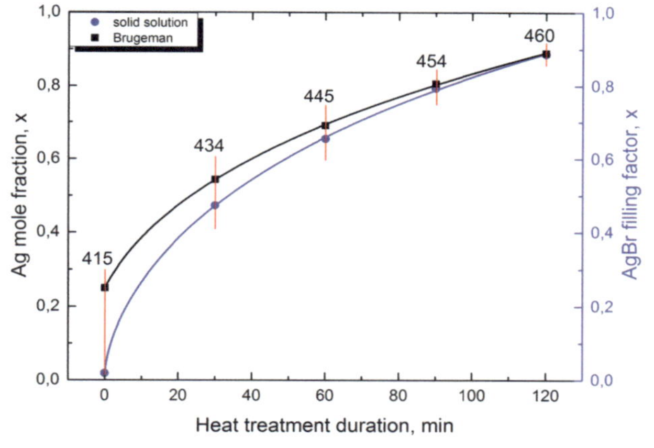

Figure 7.
Evolution of Ag mole fraction and AgBr filling factor during the heat treatment process. The corresponding SPR peak positions are denoted in nm.

If we plot the concentration of AgBr in the shell as a function of the heat treatment duration for both above cases, the plots behave monotonically (**Figure 7**) and can be approximated with sigmoidal Weibull function with zero error. This means that both models demonstrate similar behavior with rapid growth at the beginning and severe speed decrease as the system is reaching its saturation (AgBr filling factor or mole fraction).

5. Bleaching volume Bragg gratings in chloride PTR glass

In comparison with fluoride PTR glass, the chloride ones provide positive refractive index change, very high induced absorption in visible spectral range, and small size of the crystalline phase that suppresses scattering and allows recording holograms with high spatial resolution. However, the diffraction efficiency of such amplitude-phase holograms is limited according to the coupled wave theory [27]. Therefore, absorption of sinusoidal VBGs is increased and causes a decline of diffraction efficiency [28]. Thus, high absorption in near IR and visible spectral range hinders this advantage and it is highly desirable to suppress absorption of such VBGs and leave only phase grating. Since the modulation of the absorption in the grating fringes originates from the strong surface plasmon resonance of silver nanoparticles (AgNPs), which can be reduced by bleaching (photodestruction process) of the AgNPs using femtosecond laser radiation.

Figure 8 shows the optical density spectra of chloride PTR glass with recorded VBG before and after full femtosecond laser bleaching of the sample. It is seen that before the bleaching, glass possesses high absorption of surface plasmon resonance (>100 cm^{-1}) at 450 nm wavelength and visible spectral range. Inset to **Figure 9** shows the optical density spectrum of VBG after 10 hours of thermal treatment at 546°C when AgNPs concentration is fairly low. A dramatic decrease of optical density in visible can be observed after laser bleaching of the sample.

Two crystalline phases are precipitated in the volume of chloride PTR glass. The X-ray diffraction (XRD) analysis showed that the first phase is AgNPs, while the second phase is $Na_{0.9}Ag_{0.1}Cl$ (NaCl-AgCl) nanocrystals. The XRD study was performed before and after laser bleaching. The XRD analysis demonstrated the photodestruction of AgNPs playing a role of the core on silver molecular clusters and ions and preservation of NaCl-AgCl nanocrystals in the form of a shell.

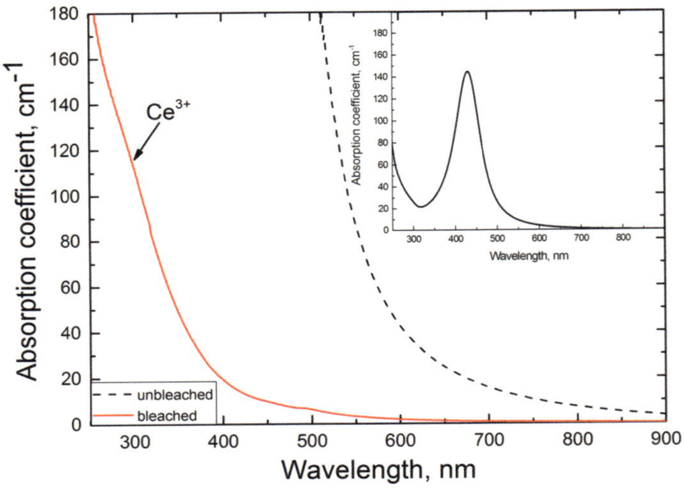

Figure 8.
Absorption spectra of as-prepared PTR glass, VBGs before and after bleaching. Inset: optical density spectrum of VBG with small AgNPs concentration.

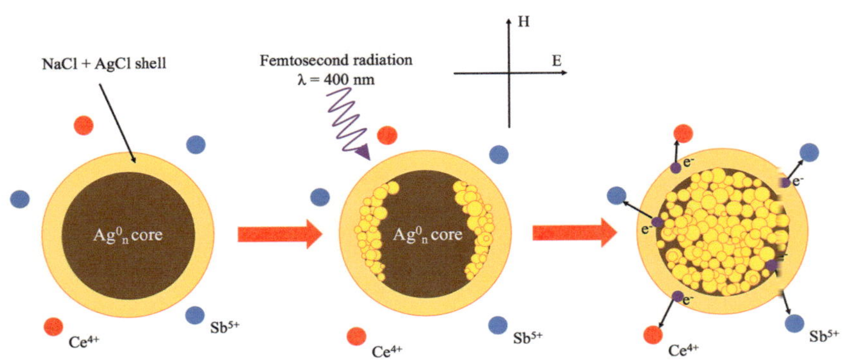

Figure 9.
The mechanism for bleaching of AgNPs hologram.

The process of photodestruction of metallic nanoparticles is well known for different types of glass [29, 30]. The absorption of ultrashort pulse energy leads to local heating and the explosion of the metallic particle [31]. In **Figure 9**, process of AgNPs surrounded by NaCl-AgCl shell, destruction, is shown schematically [32]. The femtosecond bleaching transforms AgNPs into silver clusters. However, due to the presence of shell, the particle fragments cannot move far from the origin as available space is limited by the interior volume of NaCl-AgBr crystal. It should be noted that nanoparticle restoration time under thermal treatment takes several minutes that also indicates that silver clusters close to each other after bleaching.

Considering that the absorption caused by AgNPs plasmon resonance is not present at the spectrum of the grating, it was assumed that the grating becomes purely phase type after laser processing. Previously, it was shown that VBGs in chloride PTR glass are of a mixed type, that is, both refractive and absorption indexes are modulated in the glass volume [33]. In accordance with the coupled wave theory, the coupling constant (χ) responsible for the transfer of energy between the reference (R) and signal (S) wave contains refractive index modulation amplitude (RIMA) and absorption index modulation amplitude (AIMA). Hence, removal of the one causes a natural decline of the coupling efficiency as described by Collier [34].

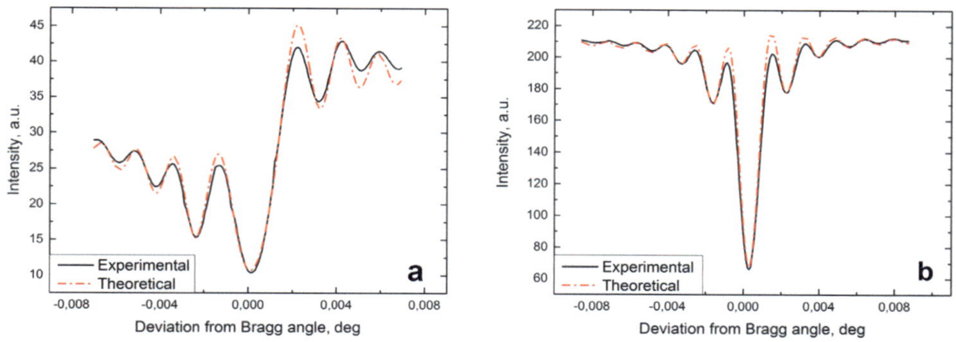

Figure 10.
Experimental and fitted curves of angular selectivity contours before (a) and after (b) photodestruction of AgNPs in VBGs.

In order to characterize the contribution of RIMA and AIMA to the VBGs performance, one should analyze angular selectivity contour of zero diffraction order. It is critical as the first-order diffraction selectivity contour looks similar for both amplitude-phase and phase gratings with the same coupling.

Experimental and fitted angular selectivity contours for the holograms recorded in chloride PTR glass are shown in **Figure 10**. One can notice an asymmetric shape of the zeroth diffraction order contour before laser bleaching with respect to Bragg angle. It directly points out at mixed nature of the grating [28, 33, 35]. After laser processing, the contour becomes symmetric as shown in **Figure 10(b)**. These factors indicate the lack of absorption modulation in the hologram, so it became purely phase. Hence, the modification of the contour shape indicates a weakening of the coupling.

The further analysis consists of the estimation of the AIMA and RIMA [27, 28, 33, 35]. The calculation of these parameters can be carried out by the fitting of the contours obtained in the experiment with those predicted by the theory (**Figure 11**, red curves). The performed fitting of the experimental results allows one to evaluate RIMA magnitude for each grating in the exposure range of 1–12 J/cm^2 before and after bleaching (**Figure 11(a)**).

As it is shown in **Figure 11(a)** for the exposures above 6 J/cm^2, the volume fraction of the crystalline phase and concentration of AgNPs saturated because RIMA value stopped increasing. Thus, all available silver and chlorine ions in the hologram fringe were utilized. Moreover, it should be noted that the RIMA values decreased after laser bleaching for all exposures and the magnitude of the change depends on the exposure.

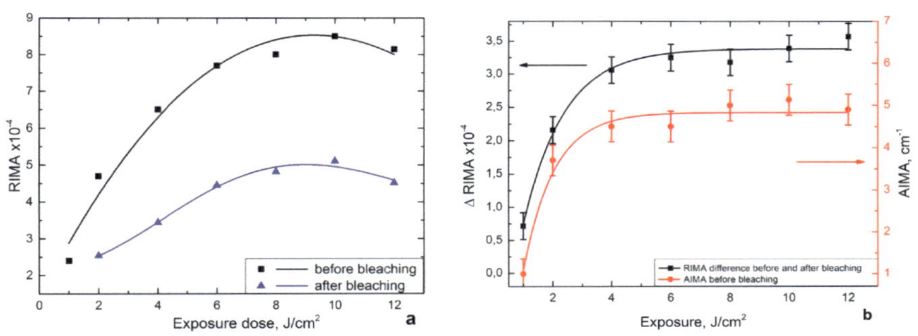

Figure 11.
(a) RIMA before and after laser bleaching of VBGs. (b) Difference of RIMA before and after laser bleaching (curve with squares) and AIMA before laser bleaching (curve with circles).

The maximum value of RIMA was reduced from 8.6×10^{-4} down to 5.1×10^{-4}, that is, it became nearly two times lower after AgNPs photodestruction. As it is following from **Figure 11(b)**, the difference in the RIMA before and after laser processing and the AIMA vs. exposure demonstrates identical dependence on the exposure. It illustrates that additional RIMA, which was lost during the laser bleaching, was caused by the presence of the AgNPs in the hologram fringes and the value is correlated with their concentration. Although the dip of the RIMA value at the maximum is considerable, it was still sufficient to increase diffraction efficiency from 25 to 86%.

Thus, ultrashort laser bleaching of chloride PTR glass with VBGs causes a decline of RIMA from 8.6×10^{-4} to 5.1×10^{-4}, which is still adequate for efficient VBGs operation. After the laser processing, the hologram becomes purely phase modulated, so it is capable to reach high diffraction efficiency at specific exposure, thermal treatment, and appropriate grating thickness. The transmission range of the VBGs expanded dramatically in visible spectral range opening avenues toward using them for high-power laser applications. Additionally, the smaller size of nanocrystals coupled with the absence of absorption in the visible region boosts the performance of such holograms at the smaller grating periods.

6. Thermal stability of volume Bragg gratings in chloride PTR glass

As-prepared chloride PTR glass is transparent in whole visible and near IR spectral range until 2.5 μm. The main drawback of chloride PTR glass is absorption in visible and near IR spectral range, caused by AgNPs possessing surface plasmon resonance (SPR) at 410–450 nm. Chloride PTR glass has high SPR absorption (>100 cm^{-1}) with considerable tail at near IR. In Section 5, we demonstrated that AgNPs absorption can be reduced using femtosecond laser radiation. Moreover, it was also demonstrated that AgNPs in chloride PTR glass can be bleached without photodestruction of NaCl-AgCl nanocrystals, responsible for refractive index modulation in VBGs. In this chapter, the bleaching effect on thermal stability of gratings recorded on chloride PTR glass is discussed. Such bleaching of chloride PTR glass with VBGs can open new opportunities for high-power laser diode (LD) applications in near-IR spectral range.

Absorption of gratings in chloride PTR glass at the wavelength of LD (972 nm) is 1.8 cm^{-1} before laser bleaching and 0.05 cm^{-1} after the bleaching. The latter corresponds to the absorption of as-prepared glass. It is clear that higher absorption of LD pumping light at wavelength 972 nm causes heat accumulation at VBG. Thus, at some point, the input power cannot be dissipated with a low thermal conductivity of the glass and the local temperature starts increasing. As following, the parameters of VBG such as refractive index difference, grating period, and absorption may change significantly.

For studying the thermal stability of VBGs, we measured the angular selectivity contour of transmission VBG at a wavelength of He-Ne laser before and after bleaching (**Figure 12**). Radiation of CW laser diode at wavelength 972 nm was used as a hitting source.

Figure 12(a) shows angular selectivity curves of VBG under different pumping of LD before femtosecond laser bleaching. A shift of Bragg wavelength under different pumping power indicates the increase of the grating period, which directly effects Bragg conditions for diffracted He-Ne laser beam. It is worth noting that the shape of the curve without laser diode irradiation is complex and contains two peaks, which indicates that the grating is overmodulated. The amplitude of central peak gradually decreases with the growth of the laser diode intensity. The actual amplitude drop is even more pronounced if one subtracts the background radiation, which can be observed far from the Bragg angle. So, it is reasonable to conclude that under high

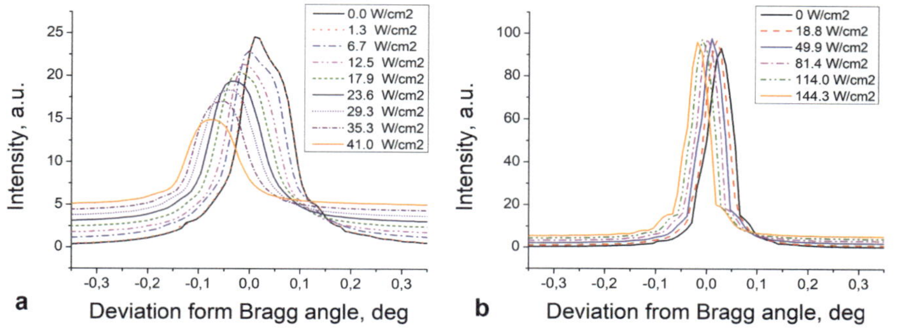

Figure 12.
Angular selectivity at first order of diffraction of initial (a) and bleached (b) hologram.

power irradiation, the diffraction efficiency drops approximately by 30%. This result can be understood by taking into account the fact that at the beam intensity of 40 W/cm^2, the grating temperature is about 350°C. It is known that the melting temperature of metal particle decreases for smaller particles. For AgNPs in silicate glass, melting temperature can be lower than 150°C, and optical properties of composite materials with embedded AgNPs change dramatically. The plasmon resonance absorption band of a liquid NP is broadening, which affects RIMA and AIMA values as it was discussed in the previous section. Melted AgNPs get crystallized back after cooling of the glass substrate with VBG and optical properties of the grating return to initial values.

Figure 12(b) shows angular selectivity contours after laser bleaching of VBGs under the different intensity of the LD diode beam. One can notice that the shift of the Bragg angle became smaller and the amplitude of the peak states unchanged even at higher fluence. The temperature increase in the irradiated area can be found from the following equation:

$$\Delta T = \frac{\Delta d}{d \cdot a} \tag{2}$$

where d is a grating period, n is a refractive index of glass, and a is a coefficient of thermal expansion. A grating temperature increase after laser bleaching at the level of 40°C for the same laser intensity is almost nine times smaller than before laser bleaching. **Figure 13** shows temperature increase and correspondent Bragg wavelength shift

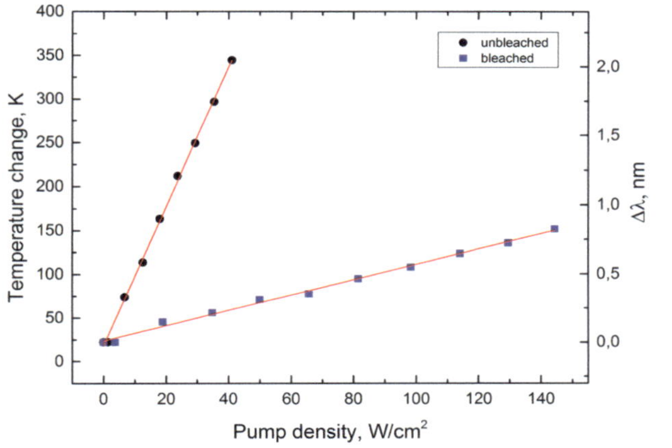

Figure 13.
Dependence of temperature change and Bragg wavelength on the power density of the pumping source.

as functions of pump irradiance for VBGs before and after laser bleaching. A dramatic decrease in temperature after laser bleaching made it possible to increase pumping intensity almost four times from 40 to 145 W/cm^2. In our experimental conditions, strong temperature gradients in the irradiated area resulted in the glass cracking when the pump intensity exceeds 145 W/cm^2.

Thus, the bleaching technology can dramatically decrease heating of the grating under high power laser radiation, which enhances the stability of the grating's parameters such as period and RIMA.

7. Conclusions

Recent achievements of ITMO University (St. Petersburg, Russia) in the investigation of properties of holographic volume Bragg gratings in fluoride and chloride photothermorefractive glass are demonstrated. Some futures of the holograms are highlighted and discussed. Some ways of improvement of characteristics of the holograms are suggested.

Namely, the mechanism of refractive index change in fluoride photothermorefractive glass during photothermoinduced crystallization and refractive index profile of the volume Bragg gratings were discussed. It is shown that refractive index change in the exposed area appears primarily due to the NaF nanocrystal inclusions in the glass matrix, which according to the Maxwell-Garnet theory decreases the effective refractive index of such composite. It is shown experimentally that residual stresses have no effect on the refractive index change. Estimated refractive index difference between the composite and unperturbed glass is in a good agreement with the refractive index change obtained by the grating analysis.

The refractive index profile in the volume Bragg gratings on fluoride photothermorefractive glass was studied. It was shown that the nonsinusoidal profile of the gratings at high exposures cannot be explained by saturation of exposure since the fraction of the first harmonic stays the same at exposures above 6 J/cm^2. In addition, it was shown that at higher exposures, grating fringes became narrower than that of the gratings with close to sinusoidal profile.

A fine structure of the core-shell system inside fluoride PTR glass in which a silver nanoparticle presents the core and crystalline phases of silver bromide and sodium fluoride present the shell was explored. In particular, for the three-layer system consisting of a silver nanoparticle, the bromide shell, and sodium fluoride shell, we made a numerical analysis of the peak location of surface plasmon resonance. The analysis showed that the thickness of the high refractive index shell does not exceed 0.5–0.6 nm. This result is in a good correlation with the unit cell size of the bromide crystals. We showed that the initial peak location of surface plasmon resonance cannot be explained by the formation of pure silver bromide shell or by the size of the silver nanoparticle. Moreover, the final peak location cannot be explained by the formation of sodium bromide shell alone. We demonstrated that the observed short-wavelength shift of the surface plasmon resonance peak at the later stage of photothermoinduced crystallization is due to the precipitation of sodium fluoride nanocrystals.

The optical properties of volume Bragg gratings in chloride PTR glass after femtosecond laser bleaching have been studied. The bleaching procedure significantly reduces the absorption of volume Bragg gratings (from 100 to 0.1 cm^{-1}). Laser bleaching does not affect the amplitude of refractive index modulation of the grating. Femtosecond laser radiation breaks down silver nanoparticle to silver clusters and ions, but the NaCl-AgCl shell remains almost the same. Thus, the amplitude-phase grating is modified in the pure phase grating by femtosecond laser photodestruction of silver nanoparticles in the volume of PTR glass. Thus, the increase of glass transmission in visible spectral range after laser bleaching of silver nanoparticles opens new opportunities for high laser power application.

It was shown that the bleaching procedure increases the thermal stability of the volume Bragg gratings under near IR laser diode pumping. So, the temperature change of the glass with volume Bragg grating under 40 W/cm^2 power density has reduced from 350 to 38°C. The bleaching of volume gratings allowed us to reduce temperature drift of Bragg wavelength from 50 to 6 pm/(W/cm^2). Thus, the bleaching technology can dramatically decrease heating of the grating under high power laser radiation, which enhances the stability of the grating parameters such as period and amplitude of refractive index modulation. It should be pointed out that the problem of silver nanoparticles absorption hinders not only chloride photothermorefractive glass because so far, all types of PTR glass (fluoride and bromide) also make use of silver nanoparticles precipitation for photothermoinduced crystallization. Thus, we believe that this approach can be utilized for bleaching of other types of PTR glass as well.

Acknowledgements

This work was supported by the Ministry of Science and Higher Education of Russian Federation (Project 16.1651.2017/4.6).

Author details

Nikonorov Nikolay*, Ivanov Sergei, Dubrovin Victor and Klyukin Dmitry
Saint-Petersburg National Research University of Information Technologies, Mechanics and Optics, St. Petersburg, Russian Federation

*Address all correspondence to: nikonorov@oi.ifmo.ru

References

[1] Stookey SD. Photosensetive glass. Industrial & Engineering Chemistry Research. 1949;**41**(4):856-861

[2] Stookey SD, Beall GH, Pierson JE. Full-color photosensitive glass. Journal of Applied Physics. 1978;**49**(10):5114

[3] Pierson JE, Stookey SD. United States Patent 4,057,408; 1977

[4] Pierson EJ, Stookey SD. United States Patent 4,017,318; 1977

[5] Panysheva EI, Tunimanova IV, Tsekhomskiĭ VA. A study of coloring in polychromatic glasses. Fizika i Khimiya Stekla. 1990;**16**(2):239-244. In Russian

[6] Dotsenko AV, Efimov AM, Zakharov VK, Panysheva EI, Tunimanova IV. On the absorption spectra of polychromatic. Fizika i Khimiya Stekla. 1985;**11**(5):592-595. In Russian

[7] Glebov LB, Nikonorov NV, Panysheva EI, Petrovskii GT, Savvin VV, Tunimanova IV, et al. New possibilities of photosensitive glasses for the recording of volume phase diagrams. Optika i Spektroskopiya. 1992;**73**(2):404-412. In Russian

[8] Kuchinskii SA, Nikonorov NV, Panysheva EI, Savvin VV, Tunimanova I. Properties of volume phase holograms on polychromatic glasses. Optika i Spektroskopiya. 1991;**70**(6):1286-1300. In Russian

[9] Nikonorov NV, Panysheva EI. Polychromatic glasses-a new medium for optical data recording. In: Opticheskoe Izobrazhenie i Registriruyushchie Sredy (All-Union Conference "Optical Image and Recording Media"). Leningrad: GOI; 1990. pp. 2-48. In Russian

[10] Glebov LB, Nikonorov NV, Panysheva EI, Tunimanova IV, Savvin VV, Tsekhomskii VA. Photothermorefractive glass. In: IF AN Latv. SSR, editor. Trudy VII Vsesoyuznoi Konferentsii po Radiatsionnoi Fizike i Khimii Neorganicheskikh Materialov (Proceedings of VII All-Union Conference on Radiation Physics and Chemistry of Inorganic Materials); Riga. 1989. p. 527. In Russian

[11] Efimov OM, Glebov LB, Glebova LN, Richardson KC, Smirnov VI. High efficiency Bragg gratings in photo-thermo-refractive glass. Applied Optics. 1999;**38**(2):619-627

[12] Glebov LB, Glebova LN, Richardson KA, Smirnov VI. Photo-induced processes in photo-thermo-refractive glasses. In: Soc AC, editor. Proceedings of XV Congress on Glass. San Francisco; 1998

[13] Ivanov SA, Ignat'ev AI, Nikonorov NV, Aseev VA. Holographic characteristics of a modified photothermorefractive glass. Journal of Optical Technology. 2014;**81**(6):356-360

[14] Nikonorov NV, Panysheva EI, Tunimanova IV, Chukharev AV. Influence of glass composition on the refractive index change upon photothermo induced crystallization. Glass Physics and Chemistry. 2001;**27**(3):241-249. Available from: http://link.springer.com/article/10.1023/A:1011392301107

[15] Glebova L, Lumeau J, Klimov M, Zanotto ED, Glebov LB. Role of bromine on the thermal and optical properties of photo-thermo-refractive glass. Journal of Non-Crystalline Solids. 2008;**354**(2-9):456-461. Available from: http://linkinghub.elsevier.com/retrieve/pii/S0022309307010903

[16] Glebov LB, Nikonorov NV, Panysheva EI, Petrovskii GT, Savvin VV, Tunimanova IV, et al. New ways to use photosensitive glasses for recording volume phase holograms. Optics and Spectroscopy. 1992;**73**(2):237-241

[17] Dyamant I, Abyzov AS, Fokin VM, Zanotto ED, Lumeau J, Glebova LN, et al. Crystal nucleation and growth kinetics of NaF in photo-thermo-refractive glass. Journal of Non-Crystalline Solids. 2013;**378**:115-120

[18] Efimov AM, Ignatiev AI, Nikonorov NV, Postnikov ES. Quantitative UV-VIS spectroscopic studies of photo-thermo-refractive glasses. I. Intrinsic, bromine-related, and impurity-related UV absorption in photo-thermo-refractive glass matrices. Journal of Non-Crystalline Solids. 2011;**357**(19-20):3500-3512

[19] Dubrovin VD, Ignatiev AI, Nikonorov NV. Chloride photo-thermo-refractive glasses. Optical Materials Express. 2016;**6**(5):1701. Available from: https://www.osapublishing.org/abstract.cfm?URI=ome-6-5-1701

[20] Sgibnev YM, Nikonorov NV, Vasilev VN, Ignatiev AI. Optical gradient waveguides in photo-thermo-refractive glass formed by ion exchange method. Journal of Lightwave Technology. 2015;**33**(17):3730-3735

[21] Nikonorov N, Ivanov S, Dubrovin V, Ignatiev A. New photo-thermo-refractive glasses for holographic optical elements: Properties and applications. In: Naydenova I, editor. Holographic Materials and Optical Systems. Dublin: InTech; 2017. pp. 435-461

[22] Lumeau J, Glebova L, Golubkov V, Zanotto ED, Glebov LB. Origin of crystallization-induced refractive index changes in photo-thermo-refractive glass. Optical Materials. 2009;**32**(1):139-146

[23] Lumeau J, Glebov LB. Effect of the refractive index change kinetics of photosensitive materials on the diffraction efficiency of reflecting Bragg gratings. Applied Optics. 2013;**52**(17):3993-3997

[24] Efimov AM, Ignatiev AI, Nikonorov NV, Postnikov ES. Quantitative UV-VIS spectroscopic studies of photo-thermo-refractive glasses. II. Manifestations of Ce^{3+} and Ce(IV) valence states in the UV absorption spectrum of cerium-doped photo-thermo-refractive matrix glasses. Journal of Non-Crystalline Solids. 2013;**361**(1):26-37

[25] Varotsos P. Determination of the dielectric constant of alkali halide mixed crystals. Physica Status Solidi. 1980;**100**(2):K133-K138

[26] Bruggeman VDAG. Berechnung verschiedener physikalischer Konstanten von heterogenen Substanzen. I. Dielektrizitätskonstanten und Leitfähigkeiten der Mischkörper aus isotropen Substanzen. Annalen der Physik. 1935;**416**(7):636-664

[27] Kogelnik H. Coupled wave theory for thick hologram gratings. Bell System Technical Journal. 1969;**48**(9):2909-2947

[28] Carretero L, Madrigal RF, Fimia A, Blaya S, Beléndez A. Study of angular responses of mixed amplitude-phase holographic gratings: Shifted Borrmann effect. Optics Letters. 2001;**26**(11): 786-788. Available from: http://www.ncbi.nlm.nih.gov/pubmed/18040450

[29] Podlipensky AV, Grebenev V, Seifert G, Graener H. Ionization and photomodification of Ag nanoparticles in soda-lime glass by 150 fs laser irradiation: A luminescence study. Journal of Luminescence. 2004;**109**:135-142

[30] Bigot J, Halte V, Merle J, Daunois A. Electron dynamics in metallic nanoparticles. Chemical Physics. 2000;**251**:181-203

[31] Hashimoto S, Werner D, Uwada T. Studies on the interaction of pulsed lasers with plasmonic gold nanoparticles toward light manipulation, heat management, and nanofabrication. Journal of Photochemistry and Photobiology C Photochemistry Reviews. 2012;**13**(1):28-54

[32] Klyukin D, Silvennoinen M, Krykova V, Svirko Y, Sidorov A, Nikonorov N. Fluorescent clusters in chloride photo-thermo-refractive glass by femtosecond laser bleaching of Ag nanoparticles. Optics Express. 2017;**25**(11):12944

[33] Ivanov SA, Nikonorov NV, Dubrovin VD, Krykova VA. Analysis of the hologram recording on the novel chloride photo-thermo-refractive glass. In: Proceedings of SPIE. The International Society for Optical Engineering; 2017

[34] Collier RJ, Burckhardt CB, Lin LH. Diffraction from volume holograms. In: Optical Holography. Amsterdam, Netherlands: Elsevier; 1971. pp. 228-264

[35] Fally M, Ellabban M, Drevenšek-Olenik I. Out-of-phase mixed holographic gratings: A qualitative analysis. Optics Express. 2008;**16**(9):6528

Chapter 2

Emulsions, Photochemistry, and Processing Factors for Display Holograms

William Alschuler

Abstract

This article reviews the range of emulsions, photochemistry, and processing techniques that have been proposed and put into practice for the successful making of display holograms. It covers various types of media including gelatin-based emulsions and photopolymers (it focuses on the former) and considers external factors that affect the final results. This is a compact review of the history of the field but focuses on the range of easily available commercial emulsions, as well as certain accounts of how to make holographic emulsions from scratch. It considers various combinations of developer, bleach, and redeveloper, which have been used to achieve the best of various trade-offs for such factors as resolution, contrast, diffraction efficiency, clarity, color quality, blackest blacks, and resistance to printout. It describes a recent advance in hypersensitizing holographic emulsions.

Keywords: holograms, holographic emulsions, hologram processing, hologram exposure factors, hologram hypersensitizing

1. Introduction

Display holography at its birth was not worth a second glance. Dennis Gabor's first images to demonstrate his new principle of image recording with scattered light were of flat objects, two-dimensional text on transparent film [1]. His idea had grown out of a desire to find a way to make an X-ray microscope, in an era when focusing X-ray optics did not exist. He conceived a way to use monoenergetic (monochromatic) X-ray scattering off objects to interfere with directly delivered X-rays of the same wavelength to create an interference pattern that coded information of the object's transparency, texture, and disposition in space, which could be recorded, and, when viewed in similar illumination, reveal a reconstructed image of the object. At that time he not only lacked focusing optics but also a medium with sufficient resolution to record the tiny-scale interference pattern details. As a demonstration he chose to use a mercury vapor discharge tube as a source, filtered to pass just one spectrum line, and photographic film to record the hologram. The limited spectral purity of the source only permitted an exposure setup with the source, filter, transparency object, and film in a straight line. This arrangement provided an interference pattern of maximum scale, one possible to record on film of ordinary resolution. The view of the hologram image seen through the hologram was a twin image centered on the light source. This is not a favorable view for display holography. However, it demonstrated that the principle was correct: an interference pattern can record details of the object, and its recording could reconstruct an image of the object when the viewing light is in the location that it occupied when the recording was made.

The game changed with the invention of lasers in 1961. Almost immediately, Leith and Upatnieks, who had been working on imaging from side-looking radar and had adopted a version of Gabor's invention to do so, realized that the laser's orders of magnitude better purity allowed an off-axis arrangement for the exposing setup and thus a view of the object without the light source mixed in [2]. They published an article with photos of the images they made with this technique, clearly demonstrating potential for realistic 3-D images that can be viewed without any optical aid [3].

Now, the production of suitable emulsions added a new requirement: it must resolve the interference pattern that, due to the change in beam geometry, became much finer, in certain respects, down to half (or less) of the wavelength of the laser light chosen. This requirement amounts to resolving at least 3000 lines per mm (lpmm) for red, 5000 lpmm for green, and at least 8,000 lpmm for blue. The grain size in the last case would need to be around 10–20 nm. These numbers contrast with resolutions of 75–150 lpmm for conventional films.

This can be done and in fact was demonstrated to be achievable back in the 1890s, by Lippmann [4, 5]. Lippmann created such an emulsion in pursuit of recording the interference pattern of ordinary light interfering with itself upon reflection at the surface of the emulsion, in contact with a mercury mirror. In principle it was the same as a reflection hologram, with full color but with no 3-D. It won him the Nobel Prize in Physics in 1908.

Ever since the work of Leith and Upatnieks, the search for emulsion materials, photochemistry, and processing has had the goals of making possible holograms with great brightness, color saturation, sharpness, contrast, clarity, minimum noise in the form of scattering, resistance to printout from further exposure to light, robustness of the emulsion in and after processing, and process safety. In these goals hides a fundamental conflict well known in ordinary photography: the inverse relationship between resolving power/grain size and sensitivity—the finer the grain, the less sensitive the emulsion. Many holographic emulsions have speeds that would lie in the range of ISO 1 or less.

It is striking that this challenge has existed in some form for the whole history of photography, the last 180 years, and that the generally best emulsion for meeting the challenge is some form of silver halide in gelatin, which has existed almost as long. Newer emulsions have come into existence that have proven useful for certain types and uses of holograms, but overall silver halides in gelatin still dominate display holography. This survey looks at some alternatives but is interested in versions of the latter, including some recent progress.

2. The range of emulsion types

In the history of holographic emulsions, silver halides in gelatin came first, as the dominating photo emulsion of the time. Then, dichromated gelatin, which was used in two old processes, carbon and gum bichromate, came into play in holography for uses that demanded the brightest possible images or greatest efficiency in focusing or diffracting light in holographic optical elements (HOEs). They are also used as master holograms for the production of mass market-reproduced holograms. The third class of materials, which includes photoresists and the possibly wide range of photopolymers, came into use for HOEs and also for masters for the mass market. They are more physically robust than those in dichromated gelatin.

Unlike all the materials listed above, which have positive indices of refraction, there is a class of materials new to visible light photonics that has a negative index of refraction. This class of materials, now called metamaterials, has moderately distant roots in some research, going back as far as the nineteenth century, with seriously renewed work in the 1950s to control radiation at radio- and microwave wavelengths [6]. The materials are artificially constructed patterned dielectrics and mixed metals and dielectrics, with structural scale at or smaller than the wavelength in question. They have been constructed to investigate various possibilities that can be envisioned that are not open to ordinary, positive-index materials. The optical properties depend on the shape and spacing of the elements of the pattern, more than the intrinsic optical properties of the materials they are constructed of. Best known of the optical element ideas is the "invisibility cloak" (think of Harry Potter's invisibility cloak), a material that can bend light so that it hides any light from what it encloses and routes the light from the background around to the front. This has been demonstrated for single wavelengths, but it is likely to be a long time before such a material can be made that creates this functionality across the whole visible spectrum. Even with this single-frequency embodiment, such a cloak could have been used in evading detection at specified wavelengths from radar-like devices.

Nonetheless, work proceeds on creating metamaterial holograms. These have so far been the versions of HOEs rather than images of ordinary subjects, because the patterns are computed before being deposited on a surface, and such things as lenses and mirrors and diffraction gratings are both computationally and depositionally simple. Groups at Duke University have designed and made holograms that store two images in the same volume, independently viewable by illuminating the hologram with light polarized at the orientation the pattern was designed for [7]. A group at the University of Pennsylvania reports constructing a flexible metamaterial made of polydimethylsiloxane (PDMS) with embedded gold rods set in precise patterns. Three images were embedded by designing three patterns, which become separately viewable as the sheet is stretched and illuminated [8].

2.1 Information recording types and diffraction efficiencies

There are three types of information storage in holograms made in original exposures in photosensitive media. If the information is recorded as opacity and transparency patterns, this is termed an amplitude recording. If the information is in the form of patterns of varying index of refraction, this is termed a phase recording. Third, the information may be stored as microscopic surface relief patterns.

Diffraction efficiency, the ratio of the energy provided in the image to the energy incident on the hologram at its viewing, is a primary determinant of the image brightness. It has been shown theoretically that phase-only holograms as well as phase combined with surface relief can achieve diffraction efficiencies close to 100%. Amplitude holograms have been shown to be limited in diffraction efficiency to about 50%. Practice over the years, in the various media types and recording schemes, shows these theoretical estimates to be correct [9].

Copying holograms for small editions or mass production faces the same limits.

2.2 Photopolymer, photoresistive, and photothermoplastic emulsions

This category of materials is basically plastics, which by various mechanisms start as monomers and then respond to light exposure by changing to polymers.

2.2.1 Photoresists

These substances, rather varied in composition, upon exposure to blue or violet light become soluble or insoluble when soaked in an organic solvent. The coating rests on a substrate, and usually the coating thickness is fairly small, one to two microns. The exposures required are up to 100 mJ/cm^2 of emulsion. These materials can be used to generate master holograms for embossed replica holograms.

This type of process recalls one of the two very earliest processes of photography: Niepce's 1827 process that employed Bitumen of Judea, coated onto a copper plate, as the photosensitive emulsion. Exposure to light caused it to polymerize and harden. Where it had not been exposed, when placed in a bath of oil of lavender, the Bitumen dissolved. The result was a relief plate that could be inked and run through a printing press.

2.2.2 Photothermoplastics

These materials typically have a multilayer structure that starts with a glass substrate; then a transparent conductive thin film, usually indium oxide; then copper contacts at the perimeter of the conductive layer; then a photoconductive layer, usually 2 microns thick; and finally on top is the thermoplastic layer, generally a fraction of micron thick, say three tenths of a micron. The surface of the thermoplastic is given a positive electric charge at its exposed surface; it is exposed to light, either blue or violet. This causes charges to migrate across the surface to form the pattern of exposure. A current is passed through the conductor to heat the thermoplastic, and it responds by forming a surface relief reproduction of the hologram interference pattern. The hologram is then ready for viewing. By heating the thermoplastic again, the charge redistributes itself to become uniform, the surface relief disappears, the image disappears, and the material is ready for use again [10].

2.2.3 Photopolymers

Photopolymers generally respond to exposure to light by polymerizing, and this generally results in changes to their density and thus their index of refraction that reproduces the interference pattern, mostly inside the emulsion. Photopolymers find a range of uses outside of holography, including some cyanoacrylate glues, bonding cement for glass, and UV-curing dental cements.

Holograms on photopolymers have a number of applications, most of which are not for display holography. These applications include security enhancement on credit cards and banknotes (so far the largest part of the market), occasionally for "flash" on packaging, makeup, trading cards, and large-volume replica images. They are also used for HOEs, such as for solar illumination control, concentrators for solar heat and photovoltaic collectors, optics for head-up displays and enhanced reality glasses, selective reflectors for laser protective glasses, optical computing, high-density optical data storage, and rapid pattern recognition.

For these as for all volume holograms, the larger the thickness, the narrower the wavelength bandwidths and also the narrower the viewing angles. In these properties volume holograms act as Bragg gratings. Narrow viewing angles are useful in many situations where holograms are used in HOEs, but usually not desirable for display holograms, where a wide audience is often a goal.

The active ingredient in the film is a photosensitive monomer that polymerizes to form the recording. The other components of a typical photopolymer film include binders (polymers that act to structure the film); photoinitiators, often various sensitizing dyes, which begin the monomer polymerization; and plasticizers, which improve film flexibility.

The gross physical structure is usually in three parts: a cover sheet that can easily be peeled off, the active photopolymer layer, and a base sheet that normally remains attached after finishing the exposure. The base sheet optical properties must be paid attention to, specifically their transparency and amount of birefringence.

In use the cover sheet is first peeled off, and then the photopolymer side, which is sticky, is hand applied to a carefully cleaned glass sheet with the help of a roller. Care is taken to avoid and remove trapped air bubbles. Also, care is taken to be dust free, at least until the image is finished [11].

The photopolymerization process is usually described on a molecular scale as starting with the absorption of light by the sensitizing dyes and active monomers. Then, the dyes linked to the monomers or other initiators generate free radicals. The free radicals capture and combine the monomers into polymers. As the process proceeds, free monomers diffuse out of their original positions into the regions where the polymers are forming. Their arrival in that area and evacuation from their original sites create density variations that in turn give rise to the variations in the index of refraction that ultimately constitute the hologram [11]. For some compositions of film, there should be as short an interval of time between exposure and heat or UV treatment as possible to avoid reaction reversal and image fading.

Related to this last issue is the fact that one of a volume hologram's desirable properties is the ability to overlay multiple images in the same volume and then to view them separately. This is the basis for high-capacity holographic memory. In order to achieve the desired goal of as many stored images as possible with uniform optical quality, it is necessary to take into account the fact that each successive exposure will tend to overwrite the preceding ones, both modifying the earlier records and eventually using up all monomers and saturating the film. Generally, later records will be weaker than earlier ones. Self-absorption by earlier records will also affect laser beam intensity of later ones, also affecting the achieved diffraction efficiency. A strategy to deal with this can be devised in the form of an exposure schedule. Assuming a constant intensity per exposure, predetermined schedule ramps up the exposure in a slightly nonlinear fashion, so that the last exposure of 350 is four times the exposure of the first [12].

One of the earliest and most important commercial families of photopolymers for holography was developed by E. I. duPont. It started with HRF-700, and now there are later versions, which are stable; have an index of refraction modulation of 0.06, which is quite high; and can be made in thicknesses up to 100 microns [11].

Dupont distributed film samples to the wider holographic community, including those who made display holograms, for a number of years, but eventually changed its policy and restricted distribution to a select "approved" list, which is now basically limited to companies using the films for mass production [13].

There are safety issues that can occur with the acrylate monomers that are often used in these films and specifically in Dupont's films. After exposure the films are UV and heat cured. Though most monomer content will typically have been converted to polymers in the exposure and UV after-exposure, the latter step can generate some acrylate monomer vapors. Therefore, curing ovens should be ventilated to the outdoors. If the vapor condenses indoors on surfaces, including the skin, it will be an oily liquid and should be removed as quickly as possible. Handwashing with soap and water is recommended at every shift. This and other directives can be found in Dupont's literature on handling their films [14].

Polaroid developed and for a time marketed a high-quality photopolymer that was available in thicknesses up to 1 mm [15]. It was discontinued when Polaroid went into bankruptcy.

There are two more recent developments in photopolymers that are worth noting.

The first is a relatively new film family by Bayer, trademarked Bayfol HX. It was supplied at least initially at 16 microns thickness, requires no postexposure heat treatment, and has an index modulation of 0.030 at the exposure wavelengths of 633, 532, and 473 nm. Typical exposure energies are 18, 25, and 30 mJ, respectively [16, 17]. It has been supplied as a four-layer film. The first is a removable cover film, the second the photopolymer, the third the substrate, and the last another removable cover film. The substrates have included low-birefringence polycarbonate and polyethylene terephthalate.

The photopolymer is a two-chemistry composite. The first is the matrix precursors, which get cross-linked during manufacturing and thereby form a matrix that provides stability. The second is the photosensitive part of the film and includes the absorbers (dyes), the initiators, and the monomers. The unexposed film can be stored and shipped (in the dark) if kept at reasonably temperature-stable conditions. During exposure the image forms. After sufficient exposure to the lasers, the film is photo-cured to bleach out the absorbers and improve transparency. No heat treatment is needed. Bayer claims that with the two "independent" photochemistries, each can be separately optimized to allow for various quality requirements.

The second development is that recently Liti Holographics entered the market with photopolymer-based kits for home and school holography and also as a supplier of large holographic images for advertising. The film is marketed for home and school use as C-RT20 in 2 × 3 inch and 4 × 5 inch sizes in packets of 6–20. The commercial services also include offers to professional photographers for add-on services for portrait photography. They promise saturated color, large view zones (120°), 7 seconds of animation, and forward projecting (i.e., real) images [18–20]. This film develops as it is exposed, so that the image slowly comes into view (in early photographic terminology, it "prints out"). It requires no further processing. The fact that this film is self-developing and requires no chemical processing is a definite plus in terms of safety, especially for home and educational uses.

The active film is specified as being 16 microns thick with a 50 micron ®cover sheet and a 1.8 mm substrate. It claims diffraction efficiencies of 99% and to be panchromatic, requiring exposures of 20, 30, and 50 mJ/cm^2 at 635, 532, and 450 nm, respectively. It is optional to post expose the film to white light that will bleach out remaining monomers and improve clarity [21].

2.2.3.1 Photopolymer holographic applications to augmented reality displays

In recent reports various applications of photopolymer holograms to augmented reality displays are described. In one design a hologram is used to provide a selectively reflecting optical element for use in eyeglass-type head-up 2-D displays [22]. The hologram is constructed as a selectively reflecting hologram, designed to create a view directly in front of the eyes, using various locations and types of illumination geometries. These include straightforward down looking informational screens, such as LCDs, and also embedding the holoscreen in a larger edge-illuminated slab, so that the projected image arrives by repeated total internal reflection at the holoscreen, which then reflects it out of the slab to the eye.

In the second, a 3-D see-through screen is achieved by a conversion of Lippmann's 1908 [23] idea of an integral photograph lenslet array to a holographic version. The original idea achieved a 3-D view of a scene by photographing it through a fly's eye lens array with the recording emulsion at the back focal plane of the array. After development and reversal to a positive, it was replaced behind the array and viewed through the array. The lenslets each provided their own views of the scene, which in total gave full parallax, both horizontal and vertical, for the viewer. In Lippmann's time producing such an array was not simple, and he apparently only got one custom-made example with which he confirmed his idea.

The new version [24] converts a lens array to a photopolymer hologram of the reflection or Denisyuk type to take advantage both of the transparency of the medium, which allows see-through views of the world beyond, and of the angular and wavelength selectivity of the hologram, which is a "thick" Bragg-angle reflector that is very wavelength selective, when illuminated with ordinary incoherent light. (Transmission-type holograms also work, but are not wavelength selective.) Illuminated at the reference beam angle, the diffraction efficiency can easily be 30%, but at the same time, the transmission through the holographic array is up around 90%. Using R, G, and B lasers to record the array, it will render full-color images, assuming the wavelengths are well chosen.

When the reference beam is replaced by a projected view of information, for example, projected from a spatial light modulator, the array forms the reflected and focused image through which the world is seen. The illumination angle can be chosen to use the phase-reversed version of the hologram. This avoids the problem of reversed depth perspective in the real image (pseudoscopy) and gives an orthoscopic version in a virtual image.

Various schemes for multiplexing 2-D and 3-D images are also explored, as are schemes for enlarging the image viewing zone. To implement large screens, which run into problems of high required laser powers and large optical systems to project the array, it is suggested that the array be constructed by mosaicking smaller arrays side by side on the polymer, and the reference beam can be diverging, instead of a collimated parallel beam.

Among the various applications, it seems that in principle the screen could be very large and allow projection of 3-D images in a Pepper's Ghost arrangement for large audiences. However, there likely would be practical difficulties.

2.2.4 An exotic potential development

A holographic recording material that apparently can be switched on and off by application of an electric current is mentioned in Ref. [25] by Bob Hess. It is described as "Holographic Polymer Dispersed Liquid Crystal (HPLDC) produced by SBG Labs (formerly DigiLens, formerly Retinal Displays)." At least one use already in production is for augmented reality helmet optics according to the DigiLens website, and there it is described as a switchable Bragg grating [26]. The grating structure is synthesized according to the optical element functionality desired.

2.2.5 A very exotic, distinctive (and tasty) family of surface relief holographic media

Back in 1997, after there was publicly available experience with photopolymers and dichromated gelatin that recorded surface relief images of the hologram interference patterns that allowed development of mass replication of master holograms, an entrepreneur experimented with and eventually succeeded in impressing a surface relief hologram on the surface of chocolate bars. To do this he started with a metal shim, made from a conventional relief master hologram, and used it to stamp copies in the chocolate. He had to find the right chocolate rigidity to record and retain the patterns, as is also needed, but perhaps less difficult to achieve in making conventional plastic copies [27]. Of course, the image quality would suffer if the chocolate went through a long enough experience of high enough temperature or was pressed down upon or rubbed in handling. This market was revisited more recently and expanded to include hard candies in addition to chocolate [28].

2.3 Photochromic materials

Photochromic materials generally darken on exposure to light. They are usually responsive to blue or UV wavelengths. Their most distinctive characteristic is that the photoreaction taking place is completely reversible, so that they make possible materials for erasable holographic memories or repeat nondestructive testing using holographic interferometry. They have been demonstrated mostly in glass doped with silver halides. These materials have essentially unlimited resolution, as the photosensitive molecules are distributed evenly throughout, not as crystals or long-chain polymers or monomers. Against this advantage there is the problem that the image decays with time, lasting perhaps 10 minutes at room temperature, though it can be slowed down by refrigeration. They are also sensitive only in the blue and UV, and not to longer wavelengths. The exposure technique began with illumination by white light, which darkened the glass uniformly, then exposure to the laser light, which selectively bleached the glass where the exposure intensity was greatest. If left alone after that, the glass gradually lightened to transparency. The image could be speed-cleared and readied for re-exposure by another white light illumination. In the 1960s I experimented using photochromic glass used for darkenable sunglasses, supplied by Corning, to make holograms. This worked in that I recorded holographic images, but the extreme insensitivity to red laser light forced 5-minute exposures with an unwidened 5 mW beam. The viewing of the reconstructed image, by laser light, will also bleach out the image, though that too could be slowed by refrigerating the sample.

2.4 Dichromated gelatin emulsions

Dichromated gelatin was one of the earliest alternatives developed to silver halide gelatin and gained favor because it recorded phase holograms directly by both very large index of refraction modulations, up in the range of 0.05, and strong surface distortions, yielding very high diffraction efficiencies and very bright images. The surface distortions can be copied onto metal that in turn could make the metal shim for stamping plastic copies. The internal transparency is excellent and the scattering noise is low. The majority of the "novelty" holograms, such as glasses with holographic eyes and keychain fobs, were mastered with or actually directly sold as sealed dichromates.

Saxby [29] gives reasonably detailed instructions for coating glass plates with DCG emulsion by three different methods:

1. Pouring along the top edge of a standing plate

2. Using a Meyer Bar to draw the emulsion across the plate while it lies horizontally

3. Spin coating the plate while pouring the emulsion on it from its center out to the edges

Gelatin can be sensitized by being premixed with dichromate (ammonium dichromate, $(NH_4)_2Cr_2O_7$) or by dipping a gelatin-coated plate into a dichromate solution.

If plates are not pre-hardened before exposure, they will tend to dissolve during processing. But this is a somewhat delicate task, because if the emulsion is too hard, then the refractive index modulation and the diffraction efficiency will be suppressed, and also scattering will increase.

The natural sensitivity of dichromated emulsion is to the blue and UV. As a result the laser that has been most used is likely the argon ion laser operating at

488 nm. Dichromate can be sensitized reasonably well to longer wavelengths by the addition of dyes. Normal exposures will be up to 100+ mJ/cm^2. It is possible to drive the diffraction efficiency to 100%, but if exposure continues, saturation can drive it down again even if the index modulation increases beyond 0.05.

Dichromated gelatin is normally processed by fixing, bathing in water, then drying by immersion in successively more concentrated baths of isopropyl alcohol in water, and finishing with 100% alcohol. After processing the emulsion must be sealed to prevent further contact with moisture, to preserve the image. This is a common practice if the image is intended for display, and not for copying. In the latter case, the surface relief is reproduced in a metal coating (a "shim") and used to stamp or injection mold the copies.

One of the pioneers of the use of DCG for display holograms and optical elements, Richard Rallison, has a good reference list [30].

2.4.1 A possible alternative DCG finishing technique

An alternative mode of finishing dichromated gelatin holograms might exist. In 1905 Lippmann commented [31], referring to his interference color photographic process, which works on the principle of a reflection hologram: "the question arose in my mind whether this transitory effect of moisture could not be permanently replaced by that of a solid, stable body."

"I soaked the plate in a solution of potassium iodide instead of in pure water. On drying, the colors were visible, though feebly so. The potassium iodide thus remained in the film, unequally distributed between maxima and minima of interference. On proceeding to pour over the dry iodized film a solution of silver nitrate (20%), the colors became extremely brilliant, and, on drying the plate, lost none of their striking character." The images were so bright that when viewed in transmission, instead of reflection, the complementary colors could be seen, and they were also brilliant. It remains to be seen whether images on emulsions processed like this remain stable or fade (blacken) due to printout.

2.5 Silver halide gelatin emulsions

Silver halide emulsions are still the most favored variety for display holograms. This is because they can be easily sensitized across the visible spectrum and are more sensitive than any of the alternatives (typically requiring exposures of 1 mJ/cm^2 or less), thus permitting the use of shorter exposures and lower power, less expensive lasers. They can also be processed to give low scattering noise and high diffraction efficiency. They are happy living unsealed in normal environments, in contrast to DCG emulsions, but less mechanically robust than photopolymers.

The gold standard for consideration of silver halide gelatin techniques is still Bjelkhagen's 1993 book [9], followed closely by more recent texts [10, 32]. The complications and considerations are numerous, but the practice has encompassed a large range of materials and processing, and many approaches allow good control of results, including color control, low noise, and high diffraction efficiency.

2.5.1 Developers

There is a very long list of developers that have been tried for holography. The most successful ones have involved one or more of the historically important developers such as metol, hydroquinone, pyrogallol, amidol, and ascorbic acid.

Development occurs in two basic ways, sometimes in combination. The first way involves reduction of the silver halide crystals that contain excited electrons created by absorbing photons and that constitute the latent image specks. The developer reduces these to form metallic silver, and the specks agglomerate to form grains. This is termed chemical development and amplifies the creation speed of the silver grains by a factor over 100,000 times. The second way uses the developer to carry silver in solution to deposit on the latent specks. These form the latent image, and a fix is used to remove unexposed silver halide. The development continues to deposit more silver on the specks. This is a rather slow process but can result in very fine grains, often called colloidal, which can decrease the light scattering and photosensitivity and thus the proclivity to printout from further exposure to light. When development is complete, the result is an amplitude hologram.

The most widely used developer formula is the one formerly issued by Kodak as D-19. It has been widely used because it gives consistently good clarity and brightness with most emulsions, is easy to use and prepare, and is one of the safest of the possibilities. It is hydroquinone (8 g)-based, with metol (2 g), sodium sulfite (90 g), sodium carbonate (52.5 g), and potassium bromide (5 g) mixed in distilled water (1 liter). It also has excellent keeping qualities.

2.5.2 Fixing

It is conventional in photography to fix the developed image by dissolving all unexposed silver halide crystals, so that further exposure to light does not darken them. This is the role of "hypo," which was discovered by Sir John Herschel at the beginning of photography in 1839. However, many fixers (including that one) shrink the emulsion as they remove the halide crystals. Fixers are used in some silver halide holography practices, but for reflection holograms, this will create a color shift to the blue. This has led to strategies that omit fixer. Desensitizers can be used in place of fixers for transmission holograms [9].

2.5.3 Bleaches

To convert an amplitude hologram to a phase hologram, and thereby achieve diffraction efficiencies in the range of 70–90%, the usual tactic is to bleach the emulsion. There is a large range of bleaches too. They are principally of two types: rehalogenating and reversal. Reversal bleaches convert the silver grains to soluble complexes that are removed from the emulsion, leaving behind the unexposed silver halide crystals, which had not been removed by fixing. This leaves voids where the silver was removed and some shrinkage. For reflection holograms this leads to color shifts, as mentioned above, but has little effect on transmission holograms. Rehalogenating bleaches recombine the developed silver with the halides. If the emulsion was fixed first, this will also result in shrinkage and color shift from the fix step. If there is no fixing, this will leave the thickness unchanged. The image contrast will receive a big boost as the image becomes transparent and is converted to a phase hologram. The migration of the silver causes voids, and this leads to some scattering, which creates a milky haze that is exposure dependent. One of the various formulas that gives good results and is relatively safe is due to Ed Wesly. It consists of ferric sulfate (12 gms), disodium EDTA (12 gms), potassium bromide (30 gms), and sodium bisulfate (50 gms), to make up to 1 liter of distilled water. It keeps very well.

2.5.4 Redevelopers

The bleached emulsion can be put through developer again. One choice is to use a developer that is slow and yields smaller and more spherical silver grains, of the colloidal type referred to above. This will reduce the scattering caused by the bleach step and can yield clear images with excellent contrast and brightness. A very simple developer for this purpose can be made by dissolving 10 gms of ascorbic acid in 1 liter of distilled water. It has a long shelf life and is safe. Soaking the bleached plate in this solution while illuminating it with a white light source will gradually darken the plate to a brown-red tone, after which it can be rinsed and dried. If the white fog is still present, it can be redeveloped again. The smaller grains are also less prone to printout.

The above regime, of D-19 developer, ferric EDTA rehalogenating bleach, and ascorbic acid redeveloper, is simple and reliably gives good results for reflection and transmission holograms. Other combinations can give better results in some cases but are mostly more dangerous, and the solutions keep less well. Often, a very last step is a soak in a wetting agent, such as Photo-Flo, which can add protection against printout.

2.5.5 Hypersensitizers and color-shift treatments (pre-exposure)

There are several methods that have been developed to hypersensitize silver halide emulsions. All three enhance sensitivity without enlarging grain size or reducing resolution. These are:

1. Pre-exposure soak in water

2. Pre-exposure soak in triethanolamine (TEA)

3. Pre-exposure soak in formate

The first two are long known to holographers. The third was recently published.

Of the first two, a simple dip of 30 seconds into distilled water followed by drying will enhance sensitivity by about a factor of 2. This effect has a shelf life of perhaps a day. The pre-exposure dip into TEA solutions followed by drying likely has a shelf life of several months. It also achieves a speed gain of about a factor of 2, without affecting resolution, though careful technique is needed if streak-free results are desired.

TEA is well known to holographers primarily for its ability to shift the color of the final result in reflection holograms. This occurs because TEA swells the emulsion. The normal interference layered fringe pattern, with spacing of ½ the laser wavelength, is recorded. Then, when the TEA is rinsed out, before development, the emulsion shrinks back, and after development and drying, the layer spacing shrinks to yield image reconstruction at a shorter wavelength. The amount of swelling effect is controlled by a combination of dip time and TEA concentration in water. The range of colors accessible by this technique is from the laser color to the bluest visible color.

There is also a post-development color-shift treatment by soaking the emulsion in a swelling agent, frequently the chemical sorbitol. This will shift the color from the laser color to colors to the red of that.

The third type of hypersensitization treatment was recently published [33]. It builds on an invention by Belloni and her coworkers that employs a pre-exposure soak in a solution of formate in water. Tests showed that a dilution of 1/100 molar would give the desired effect and leave no residue on the emulsion. The emulsion is soaked for 30 seconds to a minute and then dried. It is then exposed and developed normally. However, there is a peculiarity to the way the effect works. The speed gain is maximized by delaying development by 10–15 minutes after the exposure. With that delay, the speed gain is 10×s. If the development is done immediately after exposure, the speed gain is only a factor of 5. This converts emulsions of silver halide from requiring three or four photons per latent silver halide molecule to needing only one, a quantum efficiency of 1. There is a very slight color shift to the red, which means there was a very slight swelling of the emulsion due to the treatment.

This hypersensitizer was tested only on one holographic emulsion (Slavich PFG-01), with one developer (D-19), and for a continuous exposure, though it should in theory work for all other silver halide emulsions. It was not tested in conjunction with a pulsed laser exposure. There is evidence that latent images made by pulsed exposures fade more rapidly than by continuous exposure [9]. This leaves to open a question: will the formate pretreatment be less effective for pulsed exposures due to their tendency to rapid fading, or will it be more effective as it will slow down the rate of fading? This remains to be tested.

3. Conclusions

In this review we have looked at a range of processes and media for making high-quality holograms and some of the associated history.

The last 50 years has seen the market for holograms grow in certain commercial sectors. These have exerted pressure to develop master holographic emulsions for mass market uses such as security markers and HOEs. There have been many experiments tried, and recently two photopolymers, one by Bayer and one by Liti, have appeared. They have improved stability, improved resistance to shrinkage and color shifts, and in the case of Liti are completely self-developing, in analogy to the very old photographic printing-out processes.

Metamaterials appear to be in the process of revolutionizing holography, as well as optics in general. There will be a great deal of work done in order to commercialize them, and their primary application is likely to be for HOEs.

Display holography has not seen the market growth of other applications, despite efforts by artists and entrepreneurs, and there have not been big resources devoted to them. The most recent advance is a method of presoaking silver halide emulsions with a formate solution, which hypersensitizes them by a factor of 5–10×'s. This can allow shorter exposures, making vibration less of an issue, or allow the making of larger area images using lasers of modest power output. It can be possible to have similar effects for a wide range of photochemistries.

Conflict of interest

The author declares no conflict of interest.

Author details

William Alschuler
California Institute of the Arts, Valencia, California, USA

*Address all correspondence to: walschuler@hotmail.com

References

[1] Gabor D. A new microscopic principle. Nature. 1948;**161**:777-778

[2] Leith E, Upatnieks J. New techniques in wavefront reconstruction. Journal of the Optical Society of America A. 1961;**51**:1469

[3] Leith E, Upatnieks J. Photography by laser. Scientific American. 1965;**212**(6):24-36

[4] Lippmann G. La photographie des couleurs. *Comptes Rendus de l'Académie des Sciences*. 1891;**112**(5):274-275

[5] Lippmann G. Sur la theorie de la photographie des couleurs simples et composees par la methode interferentielle. Comptes Rendus de l'Academie des Sciences. 1894;**118**(3):92-102

[6] History of Metamaterials. Available at: https://en.wikipedia.org/wiki/History_of_metamaterials [Accessed: February 2019]

[7] Meta Group, Cummer Group, Jokerst Group, and Mikkelsen Group. Holographic Metasurfaces. 2013. Available at: https://metamaterials.duke.edu/research/holographic-metasurfaces [Accessed: February 2019]

[8] Bourzac K. Stretchy Holograms made from a metamaterial. May 10, 2017. Available at: https://cen.acs.org/content/cen/articles/95/web/2017/05/Stretchy-holograms-made-metamaterial.html [Accessed: February 2019]

[9] Bjelkhagen H. Silver Halide Emulsions for Holography and Their Processing. New York: Springer Verlag; 1993

[10] Saxby G. Practical Holography. 3rd ed. Boca Raton: CRC Press; 2003. pp. 318-322 and Appendix 5. DOI 13:978-1-4200-366-3 (e-book-PDF)

[11] Kim N. Holographic applications Based on photopolymer materials. International Workshop on Photonics and Applications, Hanoi, Vietnam, April 5-8, 2004. Available at: https://pdfs.semanticscholar.org/326d/2fa8a600f15bd172bb2825bb1915b854aba8.pdf [Accessed: February 2019]

[12] Lin SH, et al. Exposure schedule for multiplexing holograms in photopolymer. In: SPIE Conference on Photorefractive Fiber and Crystal Devices: Materials, Optical Properties and Applications V, Denver, Colorado; July 1999; SPIE. Vol. 3801

[13] John FP, and seq. Dupont Photopolymer. July 25, 2006. Available at: https://holowiki.org/forum/viewtopic.php?f=30&t=5722&sid=47e38831d8f7af17eb5c67a0affb7aa6&start=10

[14] NA. Handling procedures for dupont photopolymer films. Tech. Bulletin TB-944 (revised 03/08). Available at: http://www.dupont.com/content/dam/dupont/products-and-services/electronic-and-electrical-materials/printed-circuit-board-materials/documents/DEC-Handing-Procedures-Photopolymer-Films.pdf [Accessed: February 2019]

[15] Orlic S, Eichler HJ. Optical data storage using holographic gratings. In: Driessen A, editor. Nonlinear Optics for the Information Society Proceedings of the Third Annual COST Action P2. Boston: Kluwer Academic Publishers; 2001. pp. 79-86. DOI: 10.1007/978/94-015-1267-1

[16] Bruder F-K, Flacke T, and Rolle T. The Chemistry and Physics of Bayfol® HX Film Holographic Photopolymer. September 26, 2017. Available at: https://polymers-09-00472-v2.pdf [Accessed: February 2019]

[17] NA. https://www.films.covestro.com/~/media/Product%20Center/FF/Documents/Brochures/Bayfol%20HX%20-%20Holography.ashx [Accessed: February 2019]

[18] N. A. Liti Hologram Kit. 2019. Available at: http://www.litiholo.com/hologram-kits.html

[19] Krakow G. How to Make Holograms at Home. May 6, 2005. Available at: http://www.nbcnews.com/id/7759505/ns/technology_and_science-tech_and_gadgets/t/how-make-holograms-home/#.XGgr2Px7m7M

[20] NA. Liti Commercial Images. Available at: http://www.litiholographics.com/commercial.html [Accessed: February 2019]

[21] NA. Liti Film Specifications. Available at: https://litiholo.com/Litiholo%20CRT20%20hologram%20film%20spec%20sheet.pdf [Accessed: February, 2019]

[22] Kasai I et al. A practical see-through head mounted display using a holographic element. Optical Review. 2001;**8**(4):241-244

[23] Lippmann G. Epreuves reversible donnant la sensation du relief. Journal de Physique. 1908;7:821-825

[24] Jang C et al. Recent progress in see-through three-dimensional displays using holographic optical elements [Invited]. Applied Optics. 2016;**55**(3):A71-A85

[25] Hess R. (Bob). 2004. Available at: https://holowiki.org/forum/viewtopic.php?f=30&t=2929 [Accessed: February 2019]

[26] NA. Digilens Available at: https://www.digilens.com/ [Accessed: February 2019]

[27] Hilts PJ. Where No Candy Has Gone Before: Light as the Secret Ingredient. April 2, 1997. Available at: https://www.nytimes.com/1997/04/02/garden/where-no-candy-has-gone-before-light-as-the-secret-ingredient.html [Accessed: February 2019]

[28] Doss HM. Holograms: From Credit Cards to Chocolate. August 5, 2014. Available at: http://www.physicscentral.com/explore/action/holograms.cfm [Accessed: February 2019]

[29] Saxby G. Practical Holography. 3rd ed. Boca Raton: CRC Press; 2003. pp. 323-324 and Appendix 5. DOI 13:978-1-4200-366-3 (e-book-PDF)

[30] Rallison R. The History of Dichromates. 1996. Available at: https://wasatchphotonics.com/wp-content/uploads/The-History-of-Dichromates1.pdf [Accessed: February 2019]

[31] Lippmann G. Photographs in colors from negatives by the Lippmann process. British Journal of Photography. June 30, 1905: 505 and in the Comptes Rendu de l'Academie des Sciences (CR). 1905;**140**:1508-1509

[32] Bjelkhagen H, Brotherton-Ratcliffe D. Ultra-Realistic Imaging: Advanced Techniques in Analog and Digital Colour Holography. Boca Raton: CRC Press; 2013. pp. 59-60. DOI: 13:978-1-4-2800-7 (e-book-PDF)

[33] Alschuler WR. Formate as a sensitivity enhancer of holographic emulsions. In: Proceedings Volume 10558, Practical Holography XXXII: Displays, Materials, and Applications. Vol. 105580C. 2018. doi: 10.1117/12.2286668 [Accessed February 2019]

Section 2

Optical Systems

Chapter 3

Full-Color Holographic Optical Elements for Augmented Reality Display

Hui-Ying Wu, Chang-Won Shin and Nam Kim

Abstract

Holographic optical element is widely developed in augmented reality, virtual reality, and three-dimensional display application. Especially, the full-color HOE has been studied for the wearable device in these days. In this chapter, the basic theories and the specific analysis for the full-color holographic optical element are explained. It also explores the Bragg angle shift phenomenon caused by shrinkage of the recording materials. The full-color holographic optical element with enhanced diffraction efficiencies using the optimum recording intensities for each wavelength is presented. The fabricated full-color holographic optical element can be applied in augmented reality application. In addition, we reviewed the waveguide head-mounted display system using full-color holographic optical element.

Keywords: holographic optical element (HOE), diffraction efficiency, augmented reality (AR), photopolymer

1. Introduction

The development of display technology, which combines electronic components and fine-patterning technology, is leading to the development of advanced augmented reality (AR)/virtual reality (VR) display devices that can realize the three-dimensional (3D) world. VR is that users can use electronic equipment such as helmets to see information about a computer-generated virtual environment. And AR is a technology that integrates graphics from the computer screen or mobile display into real-world environments. In other words, AR is one of the forms in which VR is combined with reality. Head-mounted display (HMD) has been widely used in VR and AR applications [1–4]. HMD is an imaging device that can be worn on the head like a pair of glasses or a helmet [5]. Typical AR products include Microsoft's HoloLens, Google Glass and Sony's Smart Glass, while VR products include Oculus Lift, HTC Vive Pro and Samsung Gear. The basic components applicable to these AR and VR devices can be divided into electronic components and optical components. The main parts of the optical components are the diffractive optical element (DOE) [6, 7] and the holographic optical element (HOE) [8, 9]. Because DOE uses dry etching to produce fine patterns, the process is more complex than HOE production using the holographic record methods, and the costs of production are higher. While the HOE can control the angles of the two recording beams to record the interference pattern of the beams on the recording materials, so it can be produced more simply than DOE production.

The HOE is the latest area of application of holography, where a lot of interest and research is being conducted, and many conventional optical elements (such as a lens, filter, or diffraction grating) are now being replaced by the HOE [10–15]. A HOE is an optical element designed to reproduce or deform wavefront recorded on a hologram to obtain the desired wavefront. The HOE is applied in many fields because they are mass-produced, cheap, and can perform multiple functions simultaneously with a single element. Recently, it has begun to be applied to holographic head up display (HUD) systems, head-mounted display (HMD) that show driving information in front of the driver's eyes without looking at the dashboard of the car [16]. In 2015, Sony released the transparent lens eyewear "SmartEyeGlass Developer Edition". It includes a CMOS image sensor, accelerometer, gyroscope, electronic compass, brightness sensor, and microphone. Also, it is equipped with holographic waveguide technology to achieve high transparency of 85% and a thin, lightweight display.

In this chapter, the waveguide HMD system using full-color HOE is reviewed. And then it explores the Bragg angle shift phenomenon caused by shrinkage of the recording materials. The full-color HOE with enhanced diffraction efficiencies using the optimum recording intensities for each wavelength will be presented. The fabricated full-color HOE can be applied in the AR application.

2. Full-color HOE for waveguide-type HMD

When the HOE is applied to the HMD system, it can reduce the size of the overall device and the system becomes considerably lighter than the conventional HMD systems. Holographic waveguide HMD based on HOE is widely studied and developed in AR application.

Piao et al. proposed a full-color HOE using a photopolymer for a waveguide-type HMD [17]. **Figure 1** shows the schematic configuration of a holographic waveguide HMD. The waveguide HMD system can be simplified by replacing the mirror or lens of the system with HOE. The first HOE diffracts the magnified image from micro-display and the diffracted image is guided inside the waveguide plate with the total internal reflection. Finally, the guided light diffracted by the second HOE projects the image to the observers.

In order to provide the full-color HOE, the optical efficiencies of the full-color HOE with different structures are measured, as shown in **Figure 2**. Here, the optical efficiency is the intensity ratio of the diffracted beam to the incident beam. When the HOE is recorded by three wavelengths simultaneously in one photopolymer, the optical efficiencies for the three wavelengths were 16, 20, and 10%, respectively. And when the HOE is recorded in the three-layer structure, the optical efficiency of the most exterior layer is too low to provide the full-color image. So the two-layer

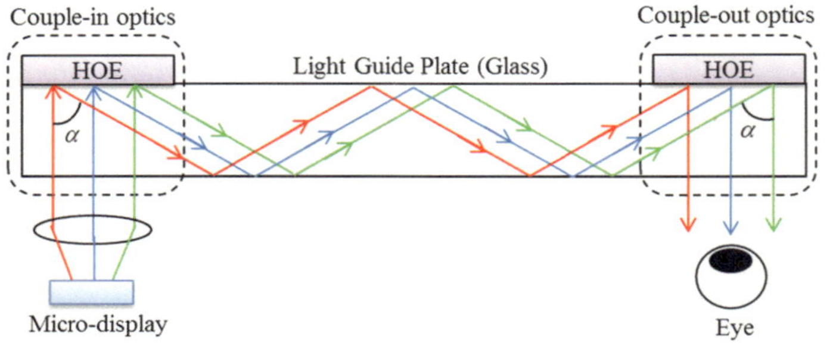

Figure 1.
The schematic configuration of a holographic waveguide HMD.

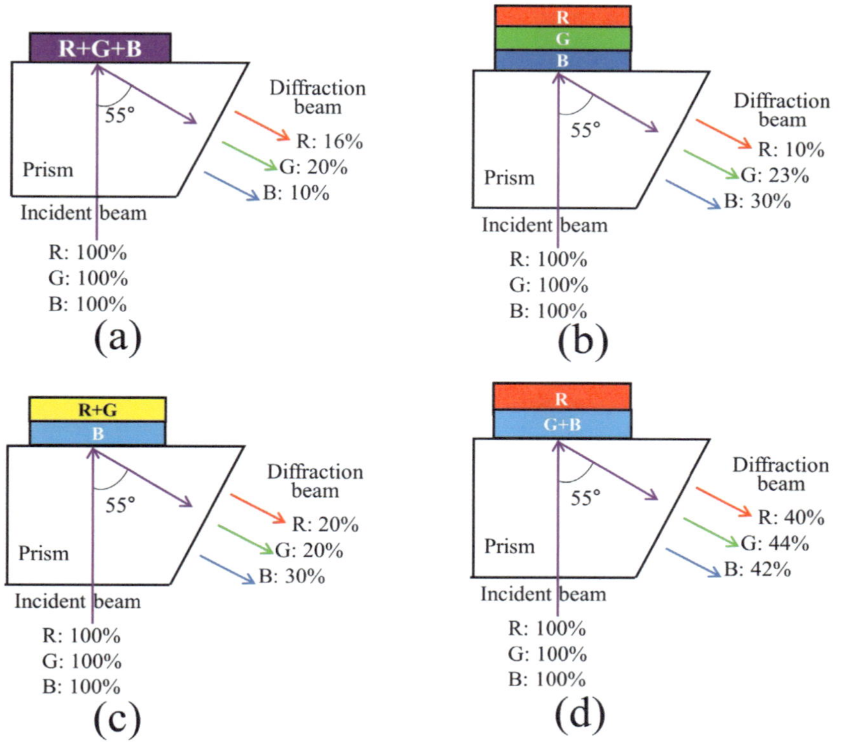

Figure 2.
The efficiency of the full-color HOEs for (a) one-layer structure, (b) three-layer structure, and (c) two-layer structure (RG/B) (d) two-layer structure (R/GB).

laminated-structure is utilized to obtain higher optical efficiency. The photopolymer has different transmittance for three wavelengths. The transmittances were measured as 85% for red, 73% for green, and 70% for blue wavelength. So two kinds of laminated-structure were considered to fabricate the full-color HOE. As shown in **Figure 2(d)**, the optical efficiencies were measured 40, 44, and 42%, respectively for each wavelength. It was much higher optical efficiency than the other two-layer structure. Thus, the laminated full-color HOEs using the two-layer structure (R/GB) were applied in waveguide HMD system, as shown in **Figure 3**. The experimental results show that the fabricated full-color HOEs can be used in the waveguide HMD system to display high-quality color images.

For reducing the thickness of the glass substrate, Piao et al. designed a wedge-shaped waveguide HMD system that achieved high levels of color uniformity and optical efficiency [18]. **Figure 4** shows the schematic configuration of a designed waveguide structure. The basic principle of the designed waveguide system is similar to the conventional waveguide system. In this system, in and out-coupled optics were designed as wedge-shaped with a certain angle. And two reflection holographic volume gratings (HVGs) are attached on each side of the wedge-shaped waveguide to guide the images generated using a micro-display to the observers. As shown in **Figure 4(b)**, the incident angles of the recording material θ_1 and θ_2 determine the angle of the total internal reflection $\theta_t = \theta_1 + \theta_2$. The angle of the total internal reflection should be larger than the critical angle of the waveguide. And the slope angle of the wedge designed as $\theta_w = 30^\circ$ is the same with θ_2. According to the analyses of the angular and spectral selectivity of the HVG [19], the incident angles $\theta_1 = 40^\circ$ and $\theta_2 = 30^\circ$ are suitable for recording the HVGs, which were attached

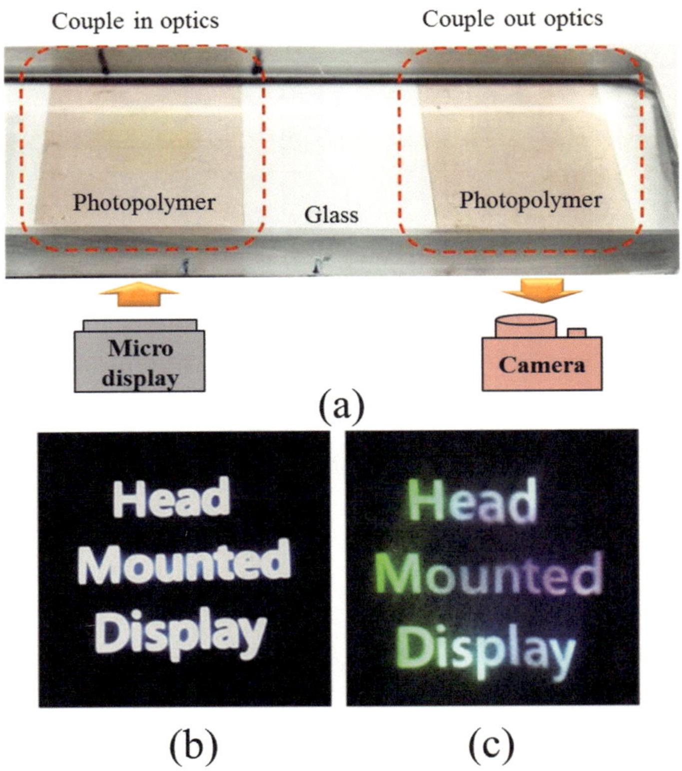

Figure 3.
(a) Holographic waveguide HMD system, (b) the original image, and (c) the guided full-color image.

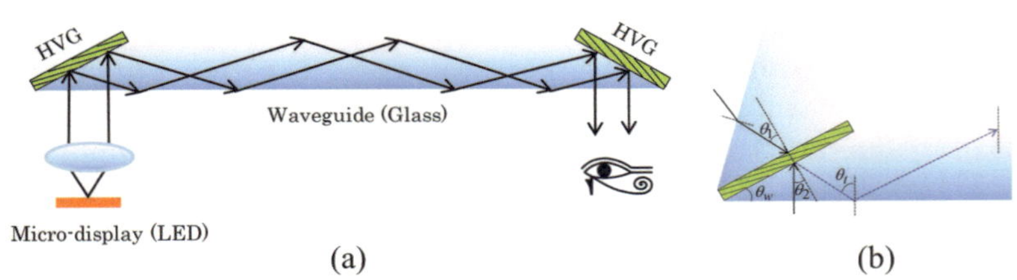

Figure 4.
(a) The schematic configuration of a wedge-shaped holographic waveguide HMD, (b) designed angle of the light path inside the wavelength.

on both sides of the wedge-shaped waveguide. The angle of the total internal reflection is directly related to the thickness of the waveguide structure. Because of the large angle of the total internal reflection, the thickness of the designed waveguide can be reduced to 1.6 times in comparison with the conventional system.

For recording the full-color HVG in one photopolymer, a multicolor reflection HVG recording method was developed for wearable display system with good color uniformity. The diffraction efficiency was improved by controlling the exposure time of each wavelength repeatedly. **Figure 5** shows the schematic of a multicolor reflection HVG recording. Shutters were used to controlling the exposure time of each wavelength.

The optical efficiencies of the multicolor HVGs with different sequence recording processes for three wavelengths are measured, as shown in **Table 1**. The results

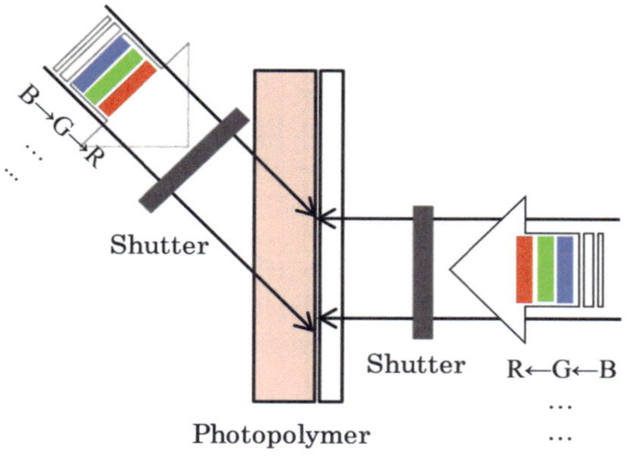

Figure 5.
Schematic of a multicolor reflection HVG recording.

Sequence of HVGs recording	Optical efficiency for each color (%)		
R G B	R (50)	G (28.6)	B (5)
R B G	R (62)	B (46)	G (31)
G R B	G (34.4)	R (51)	B (7)
G B R	G (49)	B (47)	R (44)
B R G	B (21.6)	B (59)	G (17.8)
B G R	B (15.7)	G (37.4)	R (16.4)

Table 1.
Optical efficiency of multicolor HVGs through different sequence recording processes.

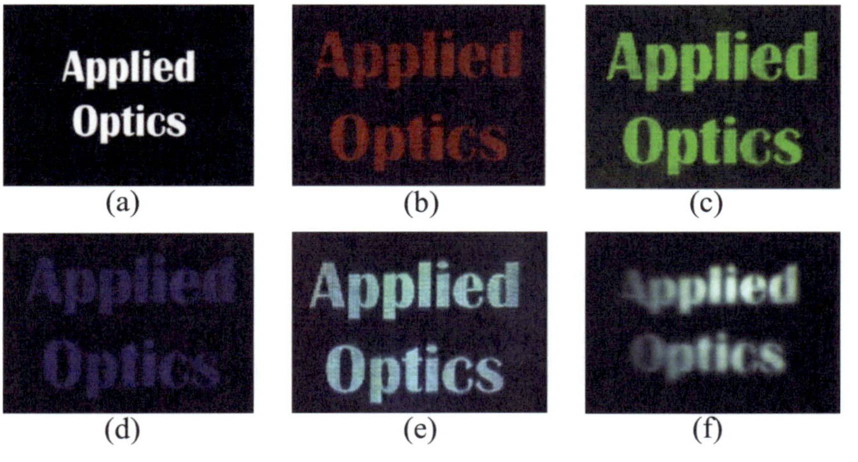

Figure 6.
Experimental results (a) original test image, output images fabricated by (b) 633 nm, (c) 532 nm, and (d) 473 nm, and (e) the GBR sequential exposure in the SLM system, (f) the GBR sequential exposure in the micro-display system.

show that the GBR recording process exhibits much higher optical efficiency and more uniform distribution than other sequential recording processes and has a high potential for obtaining high image quality.

The fabricated HVGs recorded in one photopolymer by GBR recording process were applied to the wedge-shaped waveguide system. The output images through the waveguide system were shown in **Figure 6**. The optical efficiencies of the monochromatic HVGs through the holographic waveguide system were measured as 30% for red, 34% for green, and 28% for blue wavelength. And the optical efficiency of the full-color HVG using the GBR recording process was measured 31%. **Figure 6(b)–(d)** show the output images through the wedge-shaped waveguide with each monochromatic HVGs attached. And **Figure 6(e)** shows the output image diffracted by the full-color HVGs recorded using the GBR recording process. The output image illuminated by a light emitting diode (LED) source was captured by the camera, as shown in **Figure 6(f)**. The image was clearly reproduced in white enough to be applicable to holographic waveguide HMD systems.

3. Full-color HOE for AR application

The full-color HOE, which has been studied in the past, is still thick and has some problems that need to be improved. In this section, the shrinkage compensation method was proposed to solve the Bragg angle shift phenomenon caused by the shrinkage of the recording materials in the holographic recording. And the iterative exposure time method is complicated in the recording process. So, the optimum recording intensities for each wavelength were experimentally investigated to record the three wavelengths at the same time. The fabricated full-color HOE achieved uniform diffraction efficiencies in the three wavelengths.

3.1 Measurement and compensation of shrinkage

In recording the hologram, the refraction index modulation occurs, and shrinkage of the recording materials occurs in the recording and UV heating process [20, 21]. The maximum diffraction efficiency is not measured at the designed recording angle because of the shrinkage of the recording materials, but at the shifted angle ($\Delta\theta_B$), it can reach the maximum diffraction efficiency [22].

An optical system has been implemented to compensate for the shrinkage of the recording materials. Through the optical system, it can measure the shifted angle from the Bragg angle, and then calculate the compensated value to record the hologram with the new recording angle.

Figure 7 shows a geometric configuration of the relationship between the reference beam and the signal beam for the shrinkage and compensation in the reflective asymmetric recording structure.

R is the reference beam with the incident angle θ_R, and S is the signal beam with the incident angle θ_S. R′ and S′ represent the reference beam and signal beam at the locations where the maximum diffraction efficiencies are measured due to the shrinkage of the recording materials respectively. R″ and S″ represent the reference beam and signal beam of the new calculated recording angles, considering the shrinkage compensation method. In addition, $\theta_R{'}$ is the shifted angle ($\Delta\theta_{BR}$) between the recorded reference beam R and the measured reference beam R′. $\theta_S{'}$ is the shifted angle ($\Delta\theta_{BS}$) between the recorded signal beam S and the measured signal beam S'. $\theta_R{''}$ and $\theta_S{''}$ are the calculated recording angles to compensate for the shrinkage.

The shrinkage of the recording materials is calculated by

$$d/\tan\varphi = d'/\tan\varphi' \tag{1}$$

$$Sh = \left(d - d'\right)/d \times 100\% \tag{2}$$

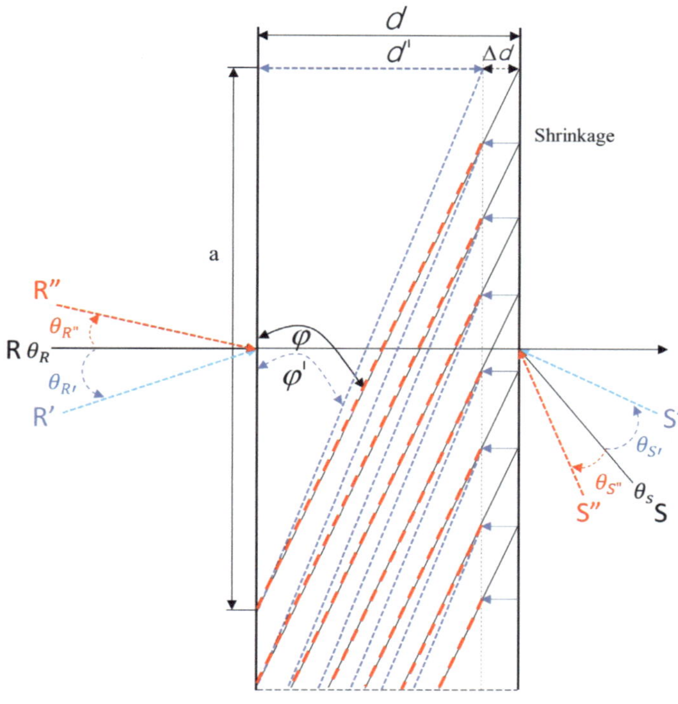

Figure 7.
Shrinkage and compensation diagram in reflective recording structure.

where d is the thickness of the material, and d′ is the thickness of the material after recording the grating. And φ is the slanted angle of the designed grating, and φ′ is the slanted angle of the recorded grating.

After recording the grating, the Bragg angle is shifted to obtain the maximum diffraction efficiency. The shifted angle $\Delta\theta_B$ can be obtained by the shrinkage of the recording materials. When the new recoding angles are calculated by $-1/2 \times \Delta\theta_B$, it can completely compensate the shrinkage to obtain the maximum diffraction efficiency at the designed recording angle.

3.2 Optimization of recording full-color HOE

The diffraction efficiency is the performance of the HOE, is measured using different exposure energies and different recording angles. When recording the HOE using a monochromatic wavelength, it's no need to concern about the effects of other wavelengths. However, when three wavelengths red, green and blue are exposed to record the grating at the same time, the response time of the recording materials vary depending on the absorption of the materials that react to each wavelength.

Three laser wavelengths, 633 nm (red) of JDSU company, 532 nm (green) of Cobolt company, 473 nm (blue) of Cobolt company are selected. And the recording material used in this section is a Bayfol HX102 photopolymer. The photopolymer has the fastest response in the red wavelength and the slowest response in the blue wavelength, so a full-color HOE should be recorded with different recording intensities.

Figure 8 shows an optical recording system diagram for recording full-color HOE. It shows the color of the laser beams for each wavelength and the color characteristics that can be expressed when laser beams are mixed with two wavelengths respectively. When recording the full-color HOE, the uniformity of the laser beam of three wavelengths will determine the color uniformity of full-color images.

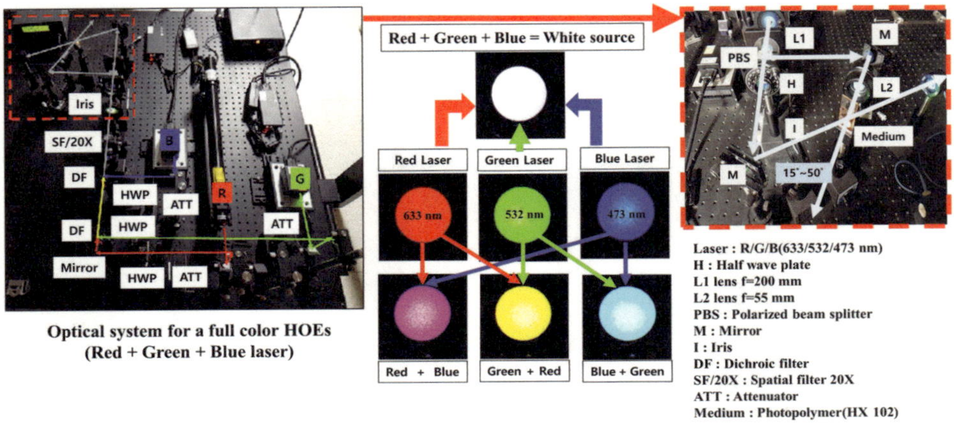

Figure 8.
Optical recording system for full-color HOE lens.

The characteristics of the photopolymer have significant effects on the many applications. Especially, the recording angles and recording intensities are very important to determine the diffraction efficiencies of the HOE. The optimized recording angle was fixed at 30° by measuring the diffraction efficiencies of different recording angles using green wavelength. Then the optical experiments were implemented corresponding to the different recording intensities for each wavelength.

If the recording intensity of each wavelength is higher than 1 mW/cm^2 when recording full-color HOE, the response time of each wavelength is so faster that the refractive index modulation reaches the saturation area quickly, causing the diffraction efficiency of each wavelength to be inconsistent or lower. And if the recording intensity is less than 0.1 mw/cm^2, before the monomer is polymerized, the photo initiators will react first, resulting in lower diffraction efficiency. Thus, through the analysis of the inhibition period of the recording material for each wavelength and the response time, the intensity of each wavelength was applied in order of blue, green and red, which allows the three wavelengths to response to the recording material at the same time.

Figure 9 shows the inhibition period for each wavelength is different corresponding to the recording intensity. Considering the recording intensity of each wavelength, the full-color HOE can be expressed more uniformly.

In **Table 2**, the highest diffraction efficiency was achieved when the optimized recording intensity for each wavelength was a red wavelength of 0.1 mW/cm^2, a green wavelength of 0.11 mW/cm^2, and a blue wavelength of 0.25 mW/cm^2, respectively.

3.3 Full-color image for HUD with HOE lens

Figure 10 shows the experimental configuration for HUD using monochromatic grating. The focal length of the collimating lens is 25 mm.

To compare the image distortion that occurs during HUD implementation using HOE, a monochromatic display (green) with a size of 6 mm × 3 mm and the brightness of 1000 cd/m^2 was used. **Figure 11** shows reconstruction images of the monochromatic grating at asymmetric recording angles and symmetric recording angles. The results of the reconstruction images show that the greater the recording angle, the more severe the image distortion caused by astigmatism. The reconstruction images of the grating recorded at the symmetric structure are better than the

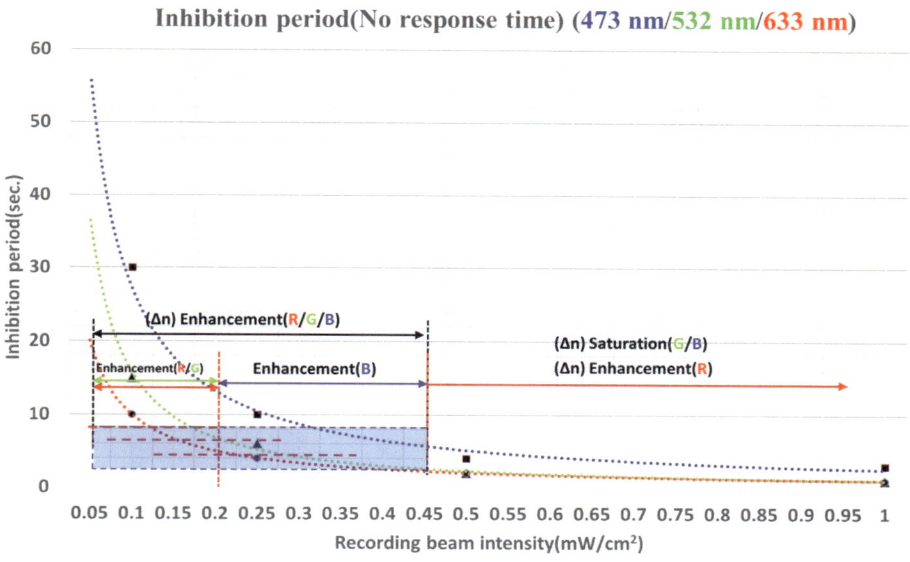

Figure 9.
Inhibition period according to each wavelength.

633 nm (mW/cm^2)	532 nm (mW/cm^2)	473 nm (mW/cm^2)	Average D. E. (%)
1	1	2.5	42.3
0.5	0.5	1	50.5
0.2	0.25	0.45	54.1
0.1	0.125	0.25	54.5
0.1	0.11	0.25	56.84
0.075	0.0825	0.1875	50.4
0.05	0.055	0.125	47.0
0.01	0.011	0.025	5.01

Table 2.
Diffraction efficiency according to recording intensities for each wavelength.

asymmetric structure. In the holographic recording, when the grating is recorded in symmetric structure with small recording angle, it can minimize the effect of astigmatism. But the diffraction efficiency of the recording material depends on the recording angle, it is important to select the optimum recording angle considering the diffraction efficiency.

Figure 12 shows a micro display spectrum compares with wavelength selectivity of red, green and blue. Because the spectral bandwidth of micro display used as HUD display differs from the wavelength selectivity bandwidth of the diffraction grating, the reflection images and diffraction images were analyzed to have large differences of color in order of red, green and blue.

Although there is a color difference, the full-color HOE can be used to express the color sufficiently, as shown in **Figure 13**. However, because the wavelength selectivity bandwidth of the recording materials is narrower than the spectral bandwidth of micro display, the visible brightness corresponds to the wavelength selectivity of the recording materials. When the spectral bandwidth of micro display

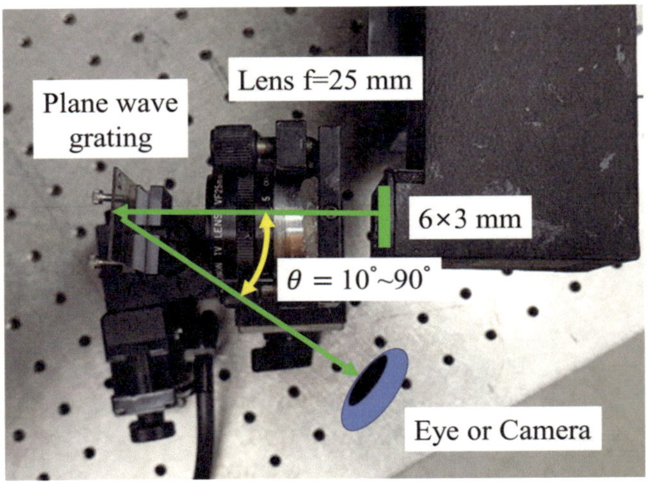

Figure 10.
Reconstruction system for monochromatic grating.

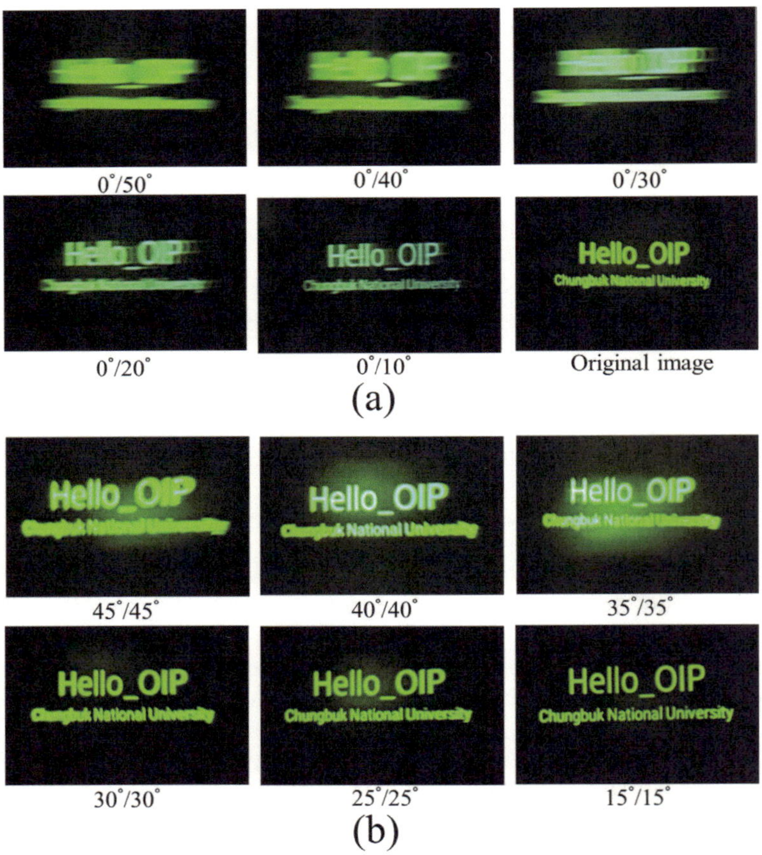

Figure 11.
Reconstruction images of monochromatic grating at (a) asymmetric recording angles (b) symmetric recording angles.

is similar to the wavelength selectivity bandwidth of the HOE, it can improve the brightness of the images displayed through the HOE. So, the spectral bandwidth of micro display can also play an important role in optimizing the full-color HOE.

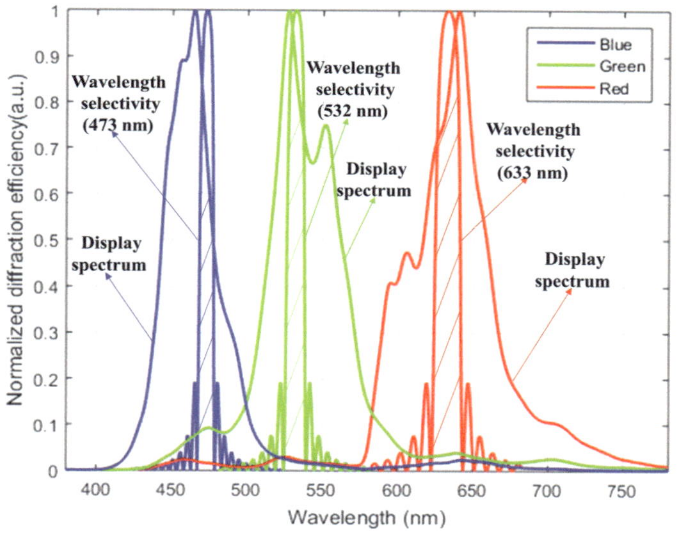

Figure 12.
Micro display spectrum and wavelength selectivity of full-color HOE.

Figure 13.
Diffraction image (left) and reflection image (right) of mixing color.

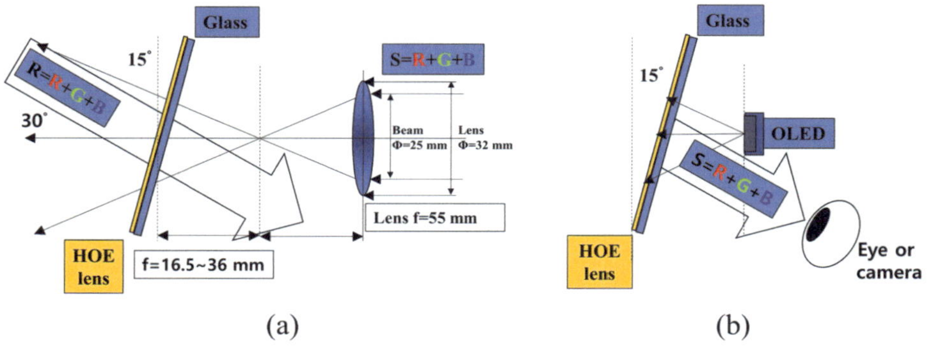

Figure 14.
Symmetric (θ = 30°) recording of full-color HOE lens and reconstruction structure (a) recording, (b) reconstruction.

Figure 14 shows the optical configuration for recording full-color HOE lenses and reconstruction. If the focal length is recorded differently, the magnification will also depend on the focal length. In other words, the functions of the optical lens are recorded in the HOE and can be used as the HOE lens.

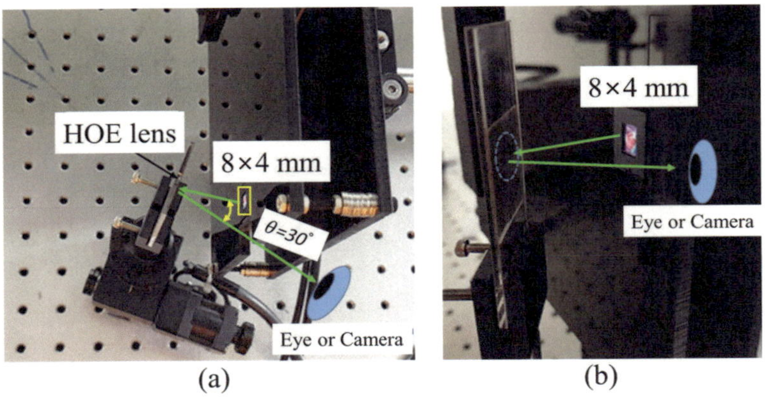

Figure 15.
Reconstruction system using full-color HOE lens (a) top view, (b) side view.

Figure 16.
Full-color reconstruction image using full-color HOE lens (a) original image, (b) reconstruction image.

The optical lens used for recording the HOE lens was utilized with two optical lenses consisting of a focal length of 55 mm with less chromatic aberration and a full diameter of 32 mm. When recording HOE lenses, the characteristics of optical lenses such as chromatic aberrations and spherical aberrations are also recorded. So optical lenses with less distortion are used.

Figure 15 shows a reconstruction system that is configured to enlarge the reconstruction image with only the HOE lens. Because the distance between the display and the HOE lens is the focal length of the HOE lens, it is possible to see the enlarged image in the eye. The size of the full-color OLED micro display is 8 mm × 4 mm, the number of the pixels is 1024 × 768, the brightness is 300 cd/m^2.

Figure 16 shows full-color images reconstructed with HOE lens using OLED micro display. The system is configured so that OLED micro display is located at the focal length of the HOE lens and can be viewed by HUD method. The reconstruction images have good color uniformity and brightness performance. The blurred images shown at the upper of the reconstruction images are images reflected by the substrate. Full-color OLED micro display can display the images by HUD method using HOE lens recorded by the optimum recording intensities for each wavelength.

4. Conclusions

The HOE is still developing in various fields such as a 3D display, holographic printer, HUD, HMD, AR and so on. Because of the lightweight, mass production advantages of the HOE, the conventional optical components are replacing with the HOE to simplify optical system. However, full-color HOE still has a limitation with color uniformity, low diffraction efficiency issues.

In this chapter, the shrinkage compensation method for the reflective diffraction grating was proposed to solve the Bragg angle shift problem. It can measure the maximum diffraction efficiency at the designed optical system without shifting the angle. And the optimum recording intensities for each wavelength were experimentally investigated by analyzing the non-response time of the recording materials. When the recording intensity was a red wavelength of 0.1 mW/cm^2, a green wavelength of 0.11 mW/cm^2, and a blue wavelength of 0.25 mW/cm^2, respectively, the diffraction efficiency of the full-color HOE reached 56.84%. The fabricated full-color HOE lens can be used in HUD systems to display the images with good color uniformity.

Acknowledgements

This work was supported in part by the National Research Foundation of Korea (NRF) grant funded by the Korea government (NRF-2017R1A2B4012096), in part by the MSIT (Ministry of Science and ICT), Korea, under the ITRC (Information Technology Research Center) support program (IITP-2019-2015-0-00448) supervised by the IITP (Institute for Information & communications Technology Promotion).

Author details

Hui-Ying Wu, Chang-Won Shin and Nam Kim*
School of Information and Communications Engineering,
Chungbuk National University, Cheongju, Chungbuk, South Korea

*Address all correspondence to: namkim@chungbuk.ac.kr

References

[1] Hua H, Girardot A, Gao C, Rolland JP. Engineering of head-mounted projective displays. Applied Optics. 2000;**39**:3814-3824

[2] Ando T, Matsumoto T, Takahashi H, Shimizu E. Head mounted display for mixed reality using holographic optical elements. Memoirs of the Faculty of Engineering. 1999;**40**:1-6

[3] Mukawa H, Akutsu K, Matsumura I, Nakano S, Yoshida T, Kuwahara M, et al. A full-color eyewear display using planar waveguides with reflection volume holograms. Journal of The Society For Information Display. 2009;**17**:185-193

[4] Yeom H-J, Kim H-J, Kim S-B, Zhang HJ, Li BN, Ji Y-M, et al. 3D holographic head mounted display using holographic optical elements with astigmatism aberration compensation. Optics express. 2015;**23**:32025-32034

[5] Tomilin MG. Head-mounted displays. Journal of Optical Technology. 1999;**66**:528-533

[6] Levola T, Aaltonen V. Near-to-eye display with diffractive exit pupil expander having chevron design. Journal of the SID. 2008;**16**:857-862

[7] Levola T. Novel Diffractive Optical Components for Near-to-Eye Displays. SID Symposium Digest of Technical Papers. Vol. 372006. pp. 64-67

[8] Oku T, Akutsu K, Kuwahara M, Yoshida T, Kato E, Aiki K, et al. High-luminance See-through Eyewear Display with Novel Volume Hologram Waveguide Technology. SID Symposium Digest of Technical Papers. 2015;**46**:192-195

[9] Kasai I, Tanijiri Y, Endo T, Ueda H. A practical see-through head mounted display using a holographic optical element. Optical Review. 2000;**8**:241-244

[10] Kim N, Piao Y-L, Wu H-Y. Holographic optical elements and application. In: Naydenova I, Nazarova D, Babeva T, editors. Holographic Materials and Optical Systems. Rijeka, Croatia: InTech; 2017. pp. 99-131. DOI: 10.5772/67297

[11] Jang G, Lee C-K, Jeong J, Li G, Lee S, Yeom J, et al. Recent progress in see-through three-dimensional displays using holographic optical elements. Applied Optics. 2016;**55**:A71-A85

[12] Hedili MK, Freeman MO, Urey H. Transmission characteristics of a bidirectional transparent screen based on reflective microlenses. Optics Express. 2013;**21**:24636-24646

[13] Wakunami K, Hsieh P-Y, Oi R, Senoh T, Sasaki H, Ichihashi Y, et al. Projection-type see-through holographic three-dimensional display. Nature Communications. 2016;**7**:12954

[14] Nakamura T, Yamaguchi M. Rapid calibration of a projection-type holographic light-field display using hierarchically upconverted binary sinusoidal patterns. Applied Optics. 2017;**56**:9520-9525

[15] Kasezawa T, Horimai H, Tabuchi H, Shimura T. Holographic window for solar power generation. Optical Review. 2016;**23**:997-1003

[16] Erdenebat M-U, Lim Y-T, Kwon K-C, Kim N. Waveguide-type head-mounted display system for AR application. In: Mohamudally N, editor. State of the Art Virtual Reality and Augmented Reality Knowhow. London, United Kingdom: InTech; 2018. pp. 41-58. DOI: 10.5772/intechopen.75172

[17] Piao J-A, Li G, Piao M-L, Kim N. Full color holographic optical element fabrication for waveguide-type head mounted display using photopolymer.

Journal of the Optical Society of Korea. 2013;**17**:242-248

[18] Piao M-L, Kim N. Achieving high levels of color uniformity and optical efficiency for a wedge-shaped waveguide head-mounted display using a photopolymer. Applied Optics. 2014;**53**:2180-2186

[19] Kogelnik H. Coupled wave theory for thick hologram gratings. Bell System Technical Journal. 1969;**48**:2909-2947

[20] Shin C-W, Vu V-T, Kim N, An J-W, Suh D, Park Y, et al. Holographic polarization-selective module based on a small Dove prism coupler for magneto-optical pickup heads. Applied Optics. 2005;**44**:4248-4254

[21] Chen JH, Su DC, Su JC. Shrinkage- and refractive-index shift-corrected volume holograms for optical interconnects. Applied Physics Letters. 2002;**81**:1387-1389

[22] Shin C-W. Holographic Optical Elements for Full Color Augmented Reality Display [thesis]. Korea: Chungbuk National University; 2019

Chapter 4

Holographic Pepper's Ghost: Upright Virtual-Image Screen Realized by Holographic Mirror

Tomoya Nakamura, Shinji Kimura, Kazuhiko Takahashi, Yuji Aburakawa, Shunsuke Takahashi, Shunsuke Igarashi, Shiho Torashima and Masahiro Yamaguchi

Abstract

A holographic mirror is a reflection-type holographic optical element that works as an off-axis mirror. It realizes an upright see-through screen serving as a virtual-image display and virtual camera. Such screen enables to realize virtual-image-based attractive applications like *Pepper's ghost* only with a thin optical system. This chapter describes the concept of a holographic-mirror-based virtual-image display and virtual camera, an experimental method for exposing the holographic mirror based on holographic printing, methods for dispersion compensation, and experimental results for the proposed virtual-image display and camera.

Keywords: holographic optical element, virtual-image display, virtual camera, dispersion compensation, volume hologram, Pepper's ghost

1. Introduction

The fusion of optical images and real objects has been an interesting topic in the field of optics and information technology. A famous example is *Pepper's ghost* [1], which was invented over 100 years ago. In *Pepper's ghost*, a virtual image is displayed on real objects by using a slanted half mirror, which can realize surprising visual experiences like optical illusions. The perception of cyber-physical fusion using virtual images mainly relies on the imperceptibility of the frame of the display, which is caused by an axial displacement between the image plane and the screen plane. Recently, such technology has been revisited in the context of augmented reality (AR). For example, virtual imaging has been used in various applications from head-mounted displays (HMDs) [2] to public theaters [3], where digital images are displayed as overlapping on real objects.

A holographic optical element (HOE) is capable of implementing various flexible optical functions on a thin, flat, and transparent film based on wavefront recording and reconstruction. Many applications of HOEs exploit their flexibility in performing optical functions and their see-through characteristics. HOEs have been applied to head-up displays (HUDs) [4], head-mount displays (HMDs) [2, 4–8], bidirectional displays [9], see-through diffusive screens [10] projection-type three-dimensional (3-D) displays [11–14], 3-D user interfaces [15], wearable

IntechOpen

eye-gaze detection systems [16], solar-power generation systems [17], vibration and temperature measurements [18, 19], and 3-D telepresence systems [20].

To realize the virtual-image-based applications only with a thin optical system, we have proposed a new optical system that integrates an HOE-based mirror referred as holographic mirror, dispersion-compensation optics, and a digital projector [21]. We also showed that a similar optical design can be applied to the realization of a virtual camera, by using a virtualization method of a camera device based on off-axis image capturing [21]. In this chapter, we describe background on the virtual-image-based applications in Section 2, a method for exposing a holographic mirror in Section 3, the concept and verification of the proposed virtual-image display in Section 4, and the concept and verification of the proposed virtual camera in Section 5. More detailed background information for the work described here is given in [21].

2. Holographic Pepper's ghost

Pepper's ghost is an illusion technique exploiting virtual images. Since the virtual image is formed outside the frame of a display, it is perceived as if it was appearing on the air. This feature is useful for realizing the unconventional visual systems based on cyber-physical fusion, which is recently referred as AR technology. This kind of visual applications can provide attractive and surprising user experiences such as the ultra-realistic telepresence system.

The classical realization of an optical system for the Pepper's ghost is based on the use of a slanted half mirror, like that in **Figure 1(a)**. It is simple to realize this arrangement of the optical system; however, the optical system will likely be bulky due to the tilted alignment. If the screen for a virtual-image display could be implemented in an upright alignment like **Figure 1(b)**, it could be integrated with flat walls, doors, windows, and existing 2-D screens. Such usage might be interesting because it allows ordinary environments to be converted into screens for virtual-image display. For instance, an ordinary wall can serve as a screen for a virtual-image-based video-communication system [22]. **Figure 2** presents the concept on such system realized by the holographic Pepper's ghost which is presented in this chapter. In the figure, a person is talking with a virtual image of another person on a real chair behind an upright window with achieving the line of sight. Exploiting the feature of the *holographic mirror*, a display for virtual-image formation can be realized by an upright thin screen unlike the conventional Pepper's ghost with a slanted half mirror.

An HOE can be used for realizing such an upright virtual-image screen. Since an HOE is a kind of hologram, flexible optical functions can be implemented on a thin

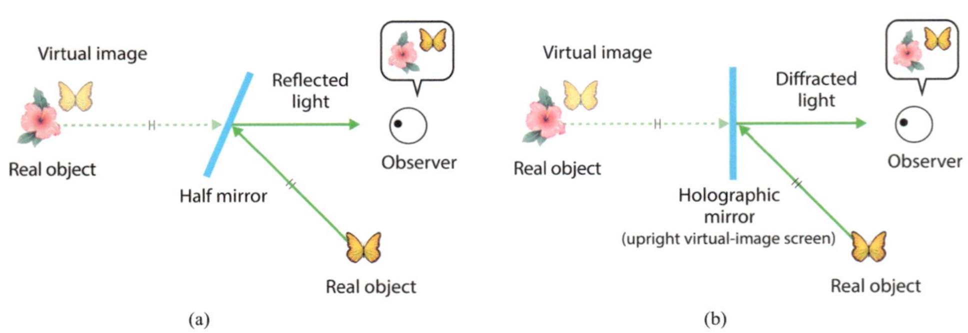

Figure 1.
Concept on (a) Pepper's ghost and (b) holographic Pepper's ghost.

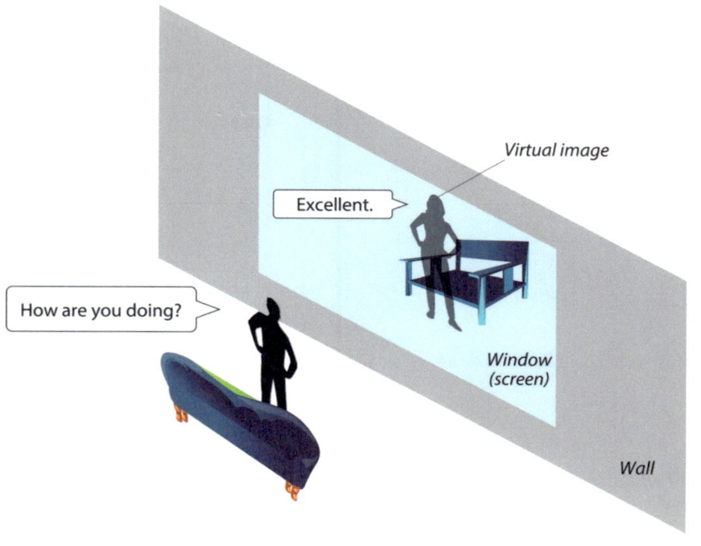

Figure 2.
An example of a virtual-image-based video-communication system based on the holographic Pepper's ghost. A person is talking with a virtual image on a real chair placed behind an upright window.

flat film by means of wavefront recording and reconstruction. For example, it is possible to realize a holographic mirror which works as an off-axis mirror by Bragg diffraction. The holographic mirror can be used for an upright virtual-image screen as mentioned above and shown in **Figure 1(b)**. One problem in applying a holographic mirror to a virtual-image screen is the chromatic dispersion caused by diffraction, which results in spatial blurring of the virtual image. This problem can simply be solved by using a laser light source; however, especially when presenting a large, deep virtual image, safety and speckle become problems. Insertion of a band-pass filter is another possible solution [16]; however, this reduces the light-use efficiency.

3. Exposure of a holographic mirror using a hologram printer

A holographic mirror can simply be implemented by exposing a photosensitive material using two coherent parallel beams. **Figure 3(a)** shows the experimental setup used in our experiment. In the setup, a diode-pumped solid-state (DPSS) laser (Samba 100 mW manufactured by Cobolt, 532 nm) was used as a light source. A half-wave plate (HWP) and a polarization beam splitter (PBS) were placed in front of the laser to split the beam with a controlled intensity ratio. In addition, an acousto-optic modulator (AOM) was inserted to function as an electrical shutter. The two beams were delivered by polarization-maintaining single-mode optical fiber (pmSMF) to regions close to the photosensitive material. We used a photopolymer (Bayfol HX200 manufactured by Covestro) as the photosensitive material. The photopolymer was exposed by the interference fringes. We adopted the scanning-based exposure method called holographic printing [23–29] to realize spatially uniform diffraction efficiency of the holographic mirror. The photopolymer was mounted on a two-axes-motorized stage for scanning.

The angle between the two beams incident on the photopolymer was set to 135°. The duration of each exposure was 10 ms, and the diameter of each beam on the photopolymer was 1.1 mm. The total number of scans was 175 × 175. The spatial interval of each scan was 0.8 mm, and the time interval between scans was 2 s. The

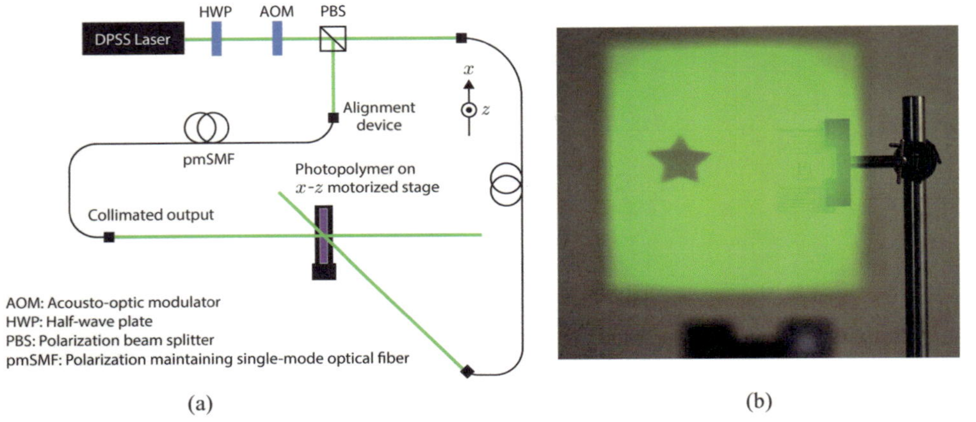

Figure 3.
(a) Experimental setup for exposing a holographic mirror and (b) the appearance of the diffracted light (a star on a spatially uniform background) by the exposed holographic mirror.

overall size of the holographic mirror was 14 ×14 cm. The appearance of a virtual image of a white paper including a star symbol formed with an exposed holographic mirror is shown in **Figure 3(b)**. Thanks to the holographic printing, the brightness of the image was spatially uniform. The diffraction efficiency measured based on ISO 17901-1 was 72.4% at 526.4 mm. More detailed information is given in [21].

4. Virtual-image display using a holographic mirror and dispersion-compensation optics

Since the holographic mirror is a volume hologram, the diffracted light should disperse chromatically, and this chromatic dispersion results in spatial blurring of the virtual image. The size of the blur caused by dispersion can be modeled as follows:

$$b = z\frac{\Delta\lambda}{\lambda}\tan\theta, \qquad (1)$$

where b is the size of the blur, z is the depth of the virtual image from the holographic mirror, $\Delta\lambda$ is the range of transmissive wavelengths through the holographic mirror, λ is the central wavelength of the propagating light, and θ is the diffraction angle of the holographic mirror. For a more generalized model, see [21]. As indicated by Eq. (1), the size of the blur linearly scales with the depth of the virtual image, the spectral width of the light, and the diffraction angle of the holographic mirror.

To suppress the blur, dispersion compensation is necessary. **Figure 4** illustrates the concept in the case of an integrated optical system with a holographic mirror, projection optics, and blur-compensation optics. The key idea is the replacement of a real display with an intentionally dispersed image, which contributes to dispersion compensation of the holographic mirror. In the system, a projector projects an image on a diffuser via a diffractive optical element (DOE). In such an optical system, a dispersed image appears on the diffuser screen. If the direction and the amount of dispersion are correctly designed for dispersion compensation, the spatial blur of the virtual image can be compensated. As a result, an observer can see a sharp virtual image through the holographic mirror.

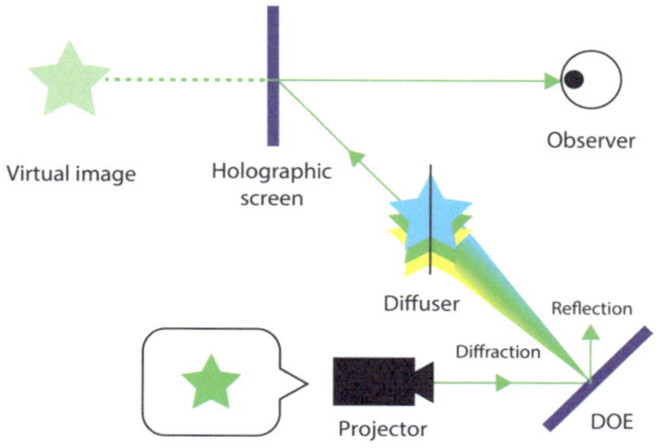

Figure 4.
Optical design of the virtual-image display with a holographic mirror and diffuser-based blur-compensation optics.

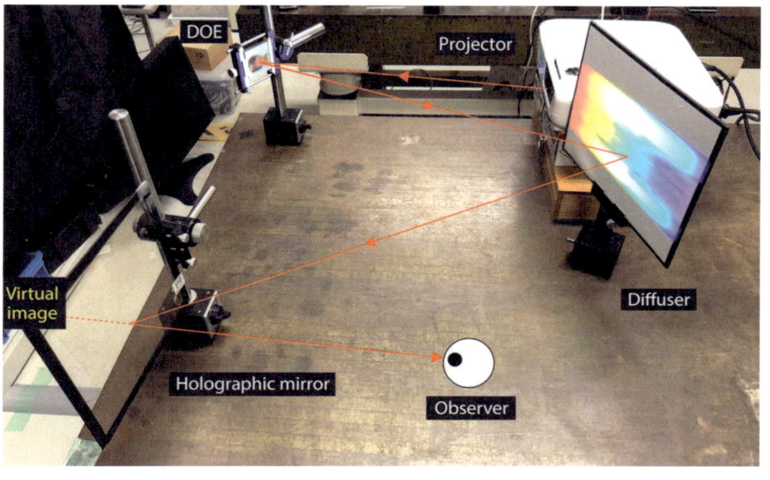

Figure 5.
Setup for experimental verification of the proposed virtual-image display.

As a related method, dispersion compensation using two identical HOEs was proposed [30, 31]. Compared with the conventional method, the advantages of the proposed DOE-based method are superior light-use efficiency and a practical level of blur suppression [21]. As mentioned also in the Introduction section, another related method is to limit the spectral width by using a laser light source or a band-pass filter. Compared with this method, our method has merits from the perspective of safety and brightness.

We verified the proposed method using the setup in **Figure 5**. We placed a reflection-type DOE (VIS Holographic Grating manufactured by Edmund optics) having 1200 grooves per millimeter in a 50 mm square in front of a projector (EH-TW5200 by EPSON) with an internal metal halide lamp. We also placed an A4-sized diffuser screen and a holographic mirror described in the previous section. We set the distance between the diffuser and the holographic mirror to 600 mm, and that between the DOE and the diffuser to 600 mm. The design conditions for the system parameters are presented in [21].

Figure 6(a) shows the images projected on the diffuser without and with a DOE. The images without the DOE were generated by replacing the DOE with a mirror. As shown in the figure, the image with the DOE was chromatically

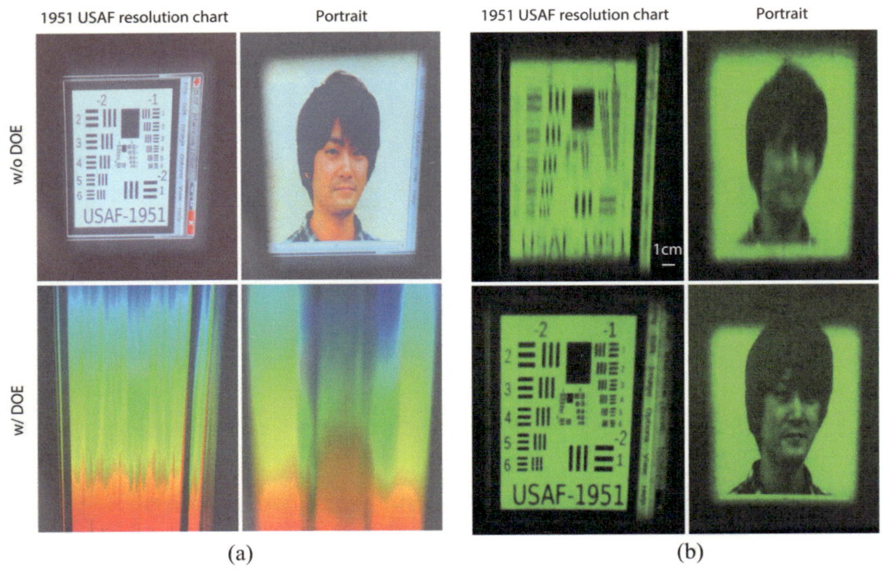

Figure 6.
(a) Projected images on a diffuser and (b) virtual images displayed by the proposed virtual-image display without and with a DOE.

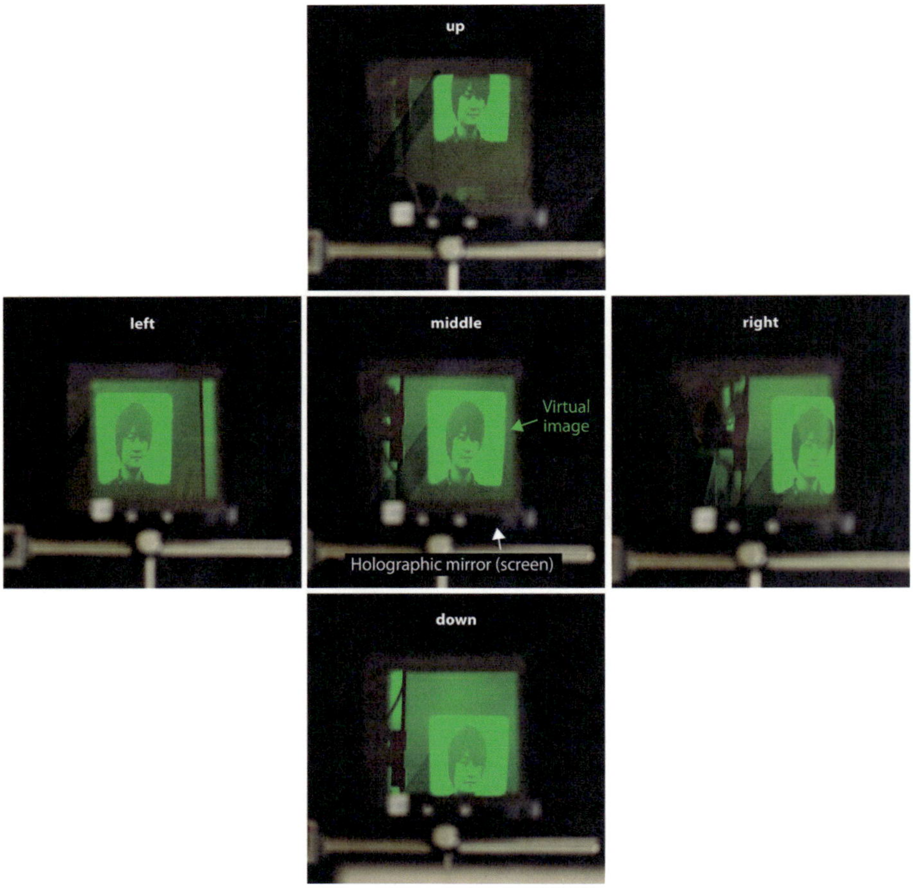

Figure 7.
Experimental virtual images observed while changing the viewing position.

dispersed along the direction of diffraction (in this case, vertical direction). The images were severely blurred in direct observation; however, these are the expected results.

Figure 6(b) shows the virtual images generated without and with a DOE, captured by a camera placed at the observer's position. The results without the DOE indicate that the virtual images were blurred along the dispersion direction. In contrast, the virtual images with the DOE were successfully resolved even along the dispersion direction. Using a resolution chart, the vertical resolution was improved from 0.12 to 0.42 cycle/mm.

Figure 7 shows the depth of the virtual images from multiple observations while changing the observer's position. As indicated in the figure, motion parallax was confirmed experimentally with a blur-compensated virtual image. In addition, since the camera focused on the virtual mirror, the holographic-mirror screen was defocused. These observations show the displacement of the axial position of the holographic mirror working as a screen and the displayed virtual image.

5. Virtual camera using holographic mirror and dispersion-compensation optics

The virtual-image display having the geometry in **Figure 1(b)** can also work inversely by replacing the projector with a camera, which can realize a *virtual camera* [21]. A virtual camera is a camera in which a virtual image of a real camera captures subjects, which allows off-axis image capturing. One benefit of off-axis image capturing is that frontal shooting of the subject can be accomplished by a camera placed at an invisible position. **Figure 8** illustrates the concept. In the figure, an observer in front of a holographic mirror is captured from the front as if an invisible camera were placed behind the mirror, where a real camera device is placed at an off-axis position. By integrating the virtual camera with the virtual-image display, a virtual-image screen that can also capture frontal images can be realized. Such a screen can be applied to, e.g., a virtual-image-based video-communication system that achieves line of sight image capturing [22].

To make use of the virtual camera with an upright holographic mirror, dispersion compensation is needed, as with the virtual display described above. In principle, the same optical system as that used for the virtual-image display can be adopted for the virtual camera; however, the insertion of a diffuser is not suitable for image capturing because the light intensity is severely reduced, and the intensity of environmental light sources (e.g., sunlight) cannot be controlled in general. To deal with this problem, we propose an alternative optical design without a diffuser

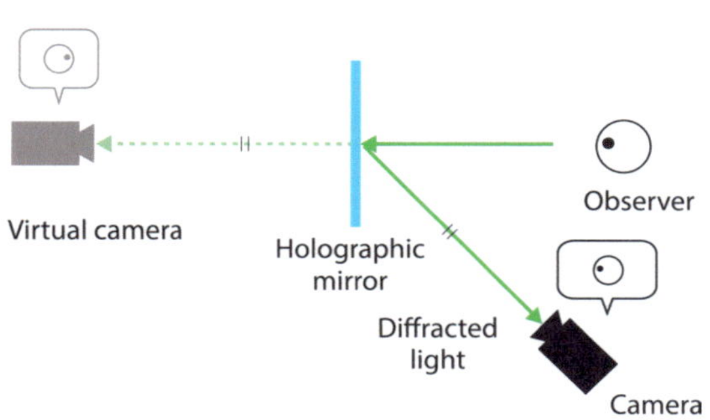

Figure 8.
Optical design for the virtual camera using a holographic mirror.

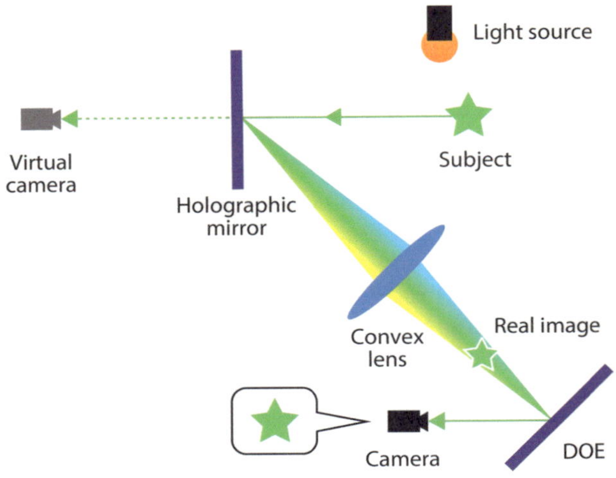

Figure 9.
Optical design of the virtual camera with a holographic mirror and the lens-based blur-compensation optics.

for dispersion compensation. **Figure 9** shows the concept of the proposed optical system. Compared with the display application in **Figure 4**, the diffuser and the projector are replaced with a convex lens and a camera, respectively. In this configuration, the light source is environmental illumination such as sunlight or room light. A convex lens is used for converting the spectrally diverging dispersed light into converging light. As a result of inserting the lens, a real image of the subject is optically formed between the lens and the DOE. A camera captures the formed real image via the DOE. Since the light is not diffused in the optical system, the light-use efficiency is superior to that of the diffuser-based display system, but on the other hand, the acceptable positions of the camera for capturing the image are restricted.

Figure 10 shows the setup used for experimental verification. The holographic mirror and the DOE are as same as those described in the previous section. The diameter and focal length of the lens were 100 and 300 mm, respectively. A color CCD camera (Flea3 manufactured by FLIR) was used for image capturing. The

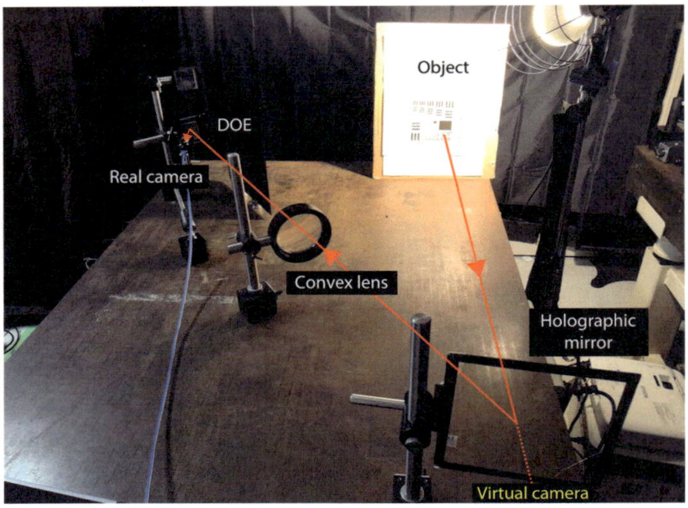

Figure 10.
Setup for experimental verification of the proposed virtual camera.

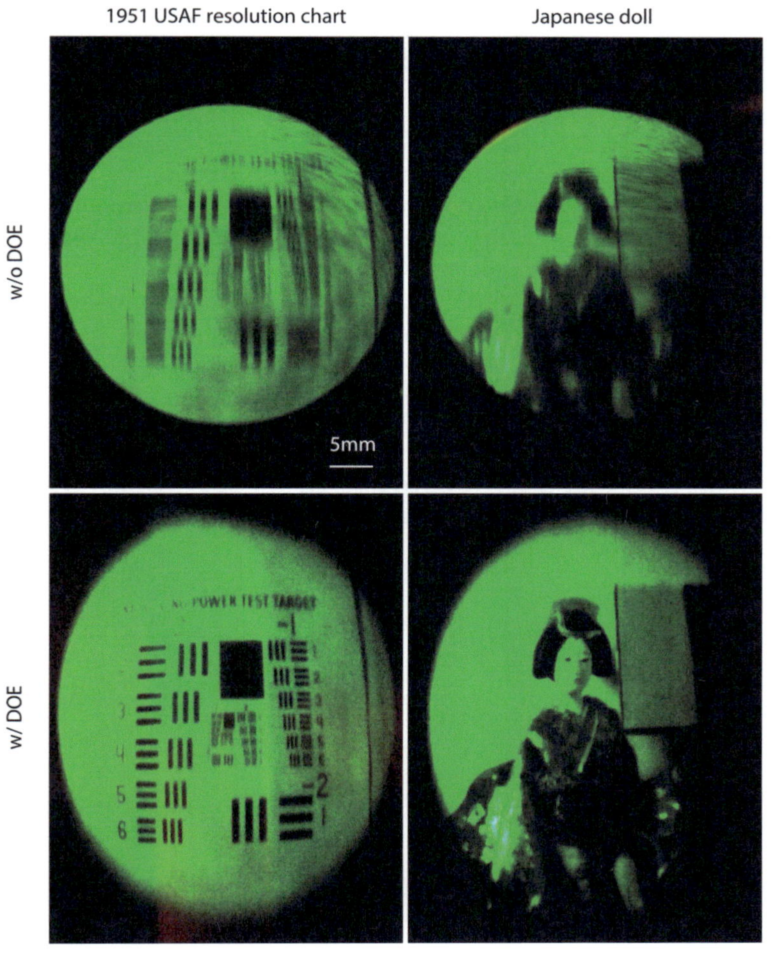

Figure 11.
Experimentally captured images obtained by the proposed virtual camera without and with a DOE.

distance between the holographic mirror and the lens was 590 mm, and that between the lens and a DOE was 680 mm. Design conditions for the system parameters are given in [21].

Figure 11 shows the images experimentally captured by the proposed virtual camera system. Without a DOE, the chromatic dispersion of the holographic mirror degraded the vertical spatial resolution of the captured image. In contrast, with the DOE, the resolution degradation was successfully restored. By visual assessment of the images of a resolution chart, the vertical spatial resolution was improved from 0.80 to 1.46 cycle/mm. The result with a doll indicates the possibility of the proposed optical system for visual video-communication systems for human users.

6. Conclusion

In this chapter, we introduced a technology on the holographic Pepper's ghost based on a virtual-image display and a virtual camera using a holographic mirror and blur-compensation optics. The holographic mirror works as an off-axis mirror, which can be used for an upright screen of the virtual-image display and the virtual camera. To make use of the holographic mirror in imaging systems, compensation of chromatic dispersion is necessary for preventing resolution degradation. We proposed two optical systems that integrate DOE-based dispersion-compensation optics, imaging devices, and a holographic mirror. In the systems, the chromatic dispersion of the holographic mirror was compensated optically. We experimentally verified the realization of the concepts on the virtual-image display and the virtual camera, and the effectiveness of the dispersion compensation.

The proposed systems can be applied to upright, thin, see-through screens for virtual-image displays and virtual cameras. The system can be used for, e.g., virtual-image-based interactive displays and video-communication systems where the screen can be integrated with environmental objects like flat walls and screen panels.

Acknowledgements

The authors would like to thank Covestro Deutschland AG for providing the photopolymer holographic recording material.

Author details

Tomoya Nakamura[1,2*], Shinji Kimura[1,3], Kazuhiko Takahashi[3], Yuji Aburakawa[3], Shunsuke Takahashi[1], Shunsuke Igarashi[1,4], Shiho Torashima[1] and Masahiro Yamaguchi[1]

1 School of Engineering, Tokyo Institute of Technology, Yokohama, Kanagawa, Japan

2 PRESTO, Japan Science and Technology Agency, Kawaguchi, Saitama, Japan

3 NTT DOCOMO, INC., Chiyoda-ku, Tokyo, Japan

4 Research Fellow of the Japan Society for the Promotion of Science, Chiyoda-ku, Tokyo, Japan

*Address all correspondence to: nakamura.t.bj@m.titech.ac.jp

References

[1] Pepper's ghost [Internet]. Available from: https://en.wikipedia.org/wiki/Pepper's_ghost [Accessed: 30-01-2019]

[2] Kasai I, Tanijiri Y, Endo T, Ueda H. A practical see-through head mounted display using a holographic optical element. Optical Review. 2001;**8**:241-244

[3] Ochi D, Kamera A, Takahashi K, Makiguchi M, Takeuchi K. VR technologies for rich sports experience. In: ACM SIGGRAPH; 21:1-21:2. 2016

[4] Withrington RJ. Optical display systems utilizing holographic lenses. UP Patent 3940204A; 1975

[5] Farrar RA. Helmet-mounted holographic aiming sight. US Patent 3633988A; 1970

[6] Mukawa H, Akutsu K, Matsumura I, Nakano S, Yoshida T, Kuwahara M, et al. A full-color eyewear display using planar waveguides with reflection volume holograms. Journal of the Society for Information Display. 2009;**17**:185-193

[7] Yeom H-J, Kim H-J, Kim S-B, Zhang H, Li B, Ji Y-M, et al. 3D holographic head mounted display using holographic optical elements with astigmatism aberration compensation. Optics Express. 2015;**23**:32025-32034

[8] Piao J-A, Li G, Piao M-L, Kim N. Full color holographic optical element fabrication for waveguide-type head mounted display using photopolymer. Journal of the Optical Society of Korea. 2013;**17**:242-248

[9] Hedili MK, Freeman MO, Urey H. Transmission characteristics of a bidirectional transparent screen based on reflective microlenses. Optics Express. 2013;**21**:24636-24646

[10] Piao M-L, Kwon K-C, Kang H-J, Lee K-Y, Kim N. Full-color holographic diffuser using time-scheduled iterative exposure. Applied Optics. 2015;**54**:5252-5259

[11] Jang G, Lee C-K, Jeong J, Li G, Lee S, Yeom J, et al. Recent progress in see-through three-dimensional displays using holographic optical elements. Applied Optics. 2016;**55**:A71-A85

[12] Wakunami K, Hsieh P-Y, Oi R, Senoh T, Sasaki H, Ichihashi Y, et al. Projection-type see-through holographic three-dimensional display. Nature Communications. 2016;**7**:12954

[13] Yamaguchi M. Light-field and holographic three-dimensional displays. Journal of the Optical Society of America A. 2016;**33**:2348-2364

[14] Nakamura T, Yamaguchi M. Rapid calibration of a projection-type holographic light-field display using hierarchically upconverted binary sinusoidal patterns. Applied Optics. 2017;**56**:9520-9525

[15] Yamaguchi M, Higashida R. 3D touchable holographic light-field display. Applied Optics. 2016;**55**:A178-A183

[16] Zhou M, Matoba O, Kitagawa Y, Takizawa Y, Matsumoto T, Ueda H, et al. Fabrication of an integrated holographic imaging element for a three-dimensional eye-gaze detection system. Applied Optics. 2010;**49**:3780-3785

[17] Kasezawa T, Horimai H, Tabuchi H, Shimura T. Holographic window for solar power generation. Optical Review. 2016;**23**:997-1003

[18] Bavigadda V, Jallapuram R, Mihaylova E, Toal V. Electronic speckle-pattern interferometer using holographic optical elements for

vibration measurements. Optics Letters. 2010;**35**:3273-3275

[19] Kumar M, Shakher C. Measurement of temperature and temperature distribution in gaseous flames by digital speckle pattern shearing interferometry using holographic optical element. Optics and Lasers in Engineering. 2015;**73**:33-39

[20] Blanche P, Bablumian A, Voorakaranam R, Christenson C, Lin W, Gu T, et al. Holographic three-dimensional telepresence using large-area photorefractive polymer. Nature. 2010;**468**:80-83

[21] Nakamura T, Kimura S, Takahashi K, Aburakawa Y, Takahashi S, Igarashi S, et al. Off-axis virtual-image display and camera by holographic mirror and blur compensation. Optics Express. 2018;**26**:24864-24880

[22] Kimura S, Nakamura T, Takahashi S, Igarashi S, Torashima S, Yamaguchi M, et al. Research of video communication system using holographic optical elements. IEICE Technical Report. 2018;**265**:21-24

[23] Yamaguchi M, Ohyama N, Honda T. Holographic three-dimensional printer: New method. Applied Optics. 1992;**31**:217-222

[24] Klug M, Holzbach M, Ferdman A. Method and apparatus for recording 1-step full-color full-parallax holographic stereograms. US Patent US6330088B1; 1998

[25] Brotherton-Ratcliffe D, Vergnes F, Rodin A, Grichine M. Holographic Printer. US Patent US7800803B2; 1999

[26] Hong K, Park S, Yeom J, Kim J, Chen N, Pyun K, et al. Resolution enhancement of holographic printer using a hogel overlapping method. Optics Express. 2013;**21**:14047-14055

[27] Yamaguchi T, Miyamoto O, Yoshikawa H. Volume hologram printer to record the wavefront of three-dimensional objects. Optical Engineering. 2012;**51**:075802

[28] Nishi W, Matsushima K. A wavefront printer using phase-only spatial light modulator for producing computer-generated volume holograms. Proceedings of SPIE. 2014;**9006**:90061F

[29] Wakunami K, Oi R, Senoh T, Sasaki H, Ichihashi Y, Yamamoto K. Wavefront printing technique with overlapping approach toward high definition holographic image reconstruction. Proceedings of SPIE. 2016;**9867**:98670J

[30] Hartman NF. Heads-up display with holographic dispersion correcting. US Patent 4613200A; 1986

[31] Klein A. Dispersion Compensation for Reflection Holography [thesis]. Cambridge: Massachusetts Institute of Technology; 1996

Chapter 5

Experimental Aspects of Holographic Projection with a Liquid-Crystal-on-Silicon Spatial Light Modulator

Michal Makowski

Abstract

Dynamic electroholography is a suitable and promising technology of image display for future projection and near-eye displays. Until a new phase modulation technology is introduced, practical research assumes the use of pixelated spatial light modulators based on liquid crystals with electronically controlled birefringence leading to a controllable refractive index. Such an approach allows for university grade development and testing of holographic computation methodology, but its limitations and drawbacks currently disable the massive application in consumer electronics. This chapter describes the differences between the behavior of the modulator as expected from Fourier optics and that observed in practical optical experiments. Moreover, practical hints and proven techniques of overcoming selected hardware issues of the chosen liquid-crystal-on-silicon (LCoS) phase modulators are given. The smart combination of the described techniques could allow more precise operation of spatial light modulators with a higher agreement with numerical simulations, especially for holographic projection of colorful images.

Keywords: holography, Fourier optics, spatial light modulator, projection, computer-generated holography

1. Introduction

Modern flat-screen liquid crystal and organic light emitting diode (OLED) displays grow in size and resolution; nevertheless, the optimal technology for even larger screens is still based on projection. The basic principle is that the real image is formed on the surface of a screen, which reflects and diffuses the intensity field toward the viewers, as it happens in cinemas and home theater systems. Image projection based on the classical optical approach utilizing lenses, high-powered lamps, and intensity modulation (e.g., DMD—digital micromirror device, DLP—digital light processing, LC—liquid crystal, or a celluloid tape) has not changed much since its invention and still has numerous disadvantages. It suffers from extremely low efficiency, which means that most of the light is converted into heat, which involves noisy, active cooling. It requires high-quality lenses to avoid visible optical aberrations—both geometrical and chromatic. Most of all, the size of the optical setup cannot be freely down-sized due to the physical requirements of the imaging process and the number and size of the apertures of the lenses. For example, shrinking the main imaging lens causes unavoidable loss of resolution due to light diffraction on the aperture of the lens. For those reasons, the image projection drifts toward the holographic image forming technique [1, 2], which is covered in this chapter.

From the practical point of view, one expects superior efficiency combined with excellent image quality and refresh rate. All those demands can be met from the theoretical point of view. The lens-less image forming by phase modulation of light with a spatial light modulator (SLM) theoretically gives 100% efficiency and allows any size of the final image, without any aberrations, noise and with real-time refresh on very large screens. Obviously, the main aspect that differ the theory from the experimental results is the hardware constraints and limitations of the SLMs currently available on the market [3]. To name a few of those aspects: the limited time response of the liquid crystal, the pixelated structure of the SLM, the non-100% fill factor, limited pixel count, lack of optical flatness, etc. [4]. Until now, better techniques of phase modulation are introduced, and one can overcome and suppress the consequences of those drawbacks of current SLMs by applying selected concepts presented in this chapter. The following subchapters will describe the experimentally validated methods of tweaking of the SLM response for the needs of good experimental realization of electro-holographic projection of 2-D images. This chapter covers only the techniques tested and experimentally proven by the author's research group. Obviously, the reader may find numerous other methods in the literature, often superior to the ones listed here.

2. The representation of a liquid crystal on silicon spatial light modulator in Fourier optics

2.1 Cartesian pixelated structure of the SLM

From the point of view of Fourier optics, the spatial light modulator provides the ability of creating a phase-only spatial 2-D field, composed of (usually square-shaped) pixels, which are associated with samples. From the mathematical point of view, the samples should contain the value of the field in their location, and moreover, they should be infinitesimally small. The pixels of popular SLMs are in the range of 8–3.74 μm and have the shape of a square. The phase retardation is the same in the whole area of the pixel, which is a major difference from the case of point-sized samples in the Fourier approach.

The nonzero size of SLM pixels greatly influences the nature of the formed optical fields, which is especially visible in their Fourier transforms. In order to illustrate that, let us assume the simplest optical realization of the Fourier transform with a converging wave illuminating the SLM, as shown in **Figure 1**.

The field reflected from the SLM carries the phase of a computer-generated hologram (CGH), therefore when the convergent beam reaches its waist, the Fourier transform of the hologram is reconstructed as an intensity field. Since the hologram was Fourier-type, the exemplary encoded image of a *Rubik cube* appears in the Fourier plane.

As seen in **Figure 2**, the hologram (signal g) displayed on the SLM must be treated as a set of samples (ergo the *combus* function) convolved with the *rect* function, which defines the shape of an individual pixel of the SLM. Therefore, the Fourier transform of such field, observed at the screen, is composed of:

- The central image (G), being an intensity Fourier transform of the input signal g.
- Copies of the G signal being the result of the convolution with the comb function.

The angular periodicity of the duplicates is governed by the law of diffraction:

$$\alpha_{diff} = \arcsin\left(\frac{m\lambda}{p}\right) \quad (1)$$

where m is the growing order numbering subsequent duplicates in a chosen direction x or y, and p is the pixel pitch of the SLM.

2.2 Pixel shape and intensity envelope

For technical reasons, the active (phase modulating) area of each pixel does not occupy 100% of the pixel's surface. The presence of residual static (non-modulating) interpixel gap accounts for the non-100% fill factor (FF) of a given SLM, defined as the ratio of the active surface of the SLM to the whole surface of

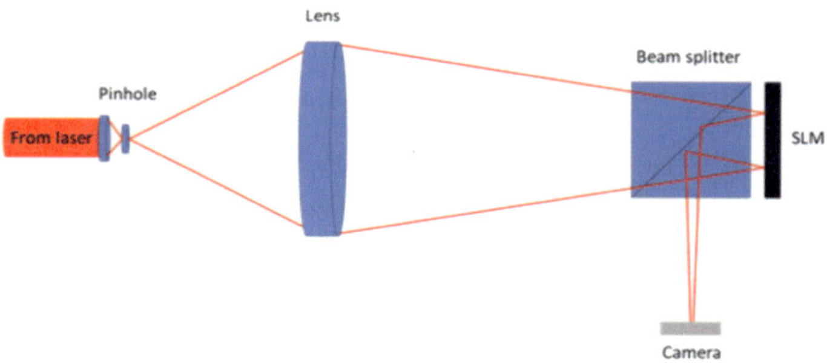

Figure 1.
Scheme of a typical optical setup for holographic projection. Only border light rays without refraction are shown for clarity.

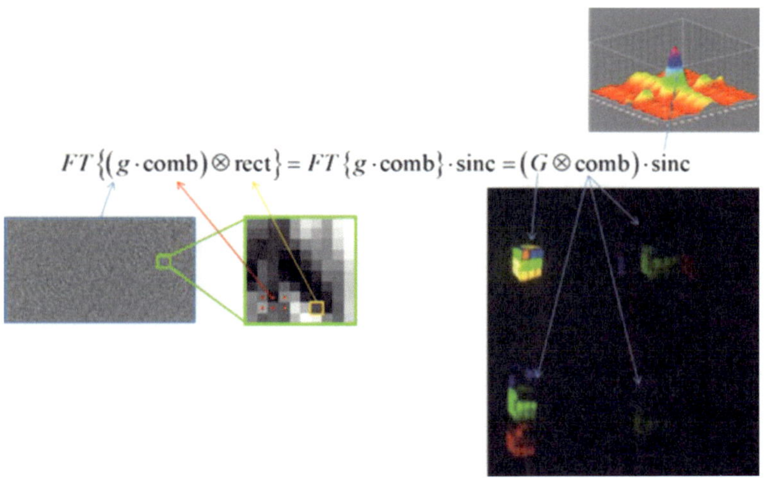

Figure 2.
Components of the Fourier transform of an input field g displayed on a pixelated SLM.

the SLM. Typically, this value is within the range of 80–95% (e.g., 87% for Holoeye Pluto-1, 93% for Holoeye Pluto-2, 90% for Holoeye GAEA) and tends to rise for lower pixel pitch values, due to minimal width of the interpixel lines in CMOS technology.

The intensity pattern visible in **Figure 2** is attenuated by the so-called intensity envelope, denoted as *sinc*. Its shape in first approximation is the Fourier transform of the shape of an individual pixel of the SLM. For square-shaped pixels, the amplitude of the envelope function is: $\mathrm{sinc}(x) = \sin(x)/x$. The attenuation of the central usable image is rather straightforward to take into account at the stage of computation of the CGH. Simply, the input image must be edited in a way that its central regions are artificially darkened so as to overexpose the boundary regions in the projection [5].

The minima of the *sinc* envelope have the position determined by the size of the *rect* function describing the active (phase modulating) part of the individual pixel, for example, 7.7 μm in **Figure 3**. For FF = 100%, the pixel pitch and the pixel size are equal, which luckily results in the minima of the attenuating function falling in the central locations of the duplicate intensity patterns, as shown in **Figure 4a**.

For any value of FF lower than 100%, the mismatch of the mentioned two functions causes the increased visibility of the spurious copies, often referred to as *higher order images*, as seen in **Figure 4b**. Therefore, the limited fill factor boosts the visibility of useless higher orders and takes the energy away from the useful first diffractive order (the central one in **Figure 2**).

2.3 Spurious orders of diffraction

The replicas of colorful holographically projected fields disappear when the angle of diffraction of the blue light in the second diffractive order reaches 90°, that is, for the pixel pitch of the SLM equal to the wavelength (e.g., 445 nm). Currently, available state of the art SLMs has the pixel pitch as large as 3.74 μm, therefore such a straightforward method of mitigation of higher orders will be the matter of distant future.

The formation of the stray higher orders of diffraction is especially problematic in holographic translucent displays, where duplicates of virtual images are created in the peripheral areas of the user's view, often with a noticeable color breakup—see **Figure 2**. They are easily perceivable and difficult to obstruct

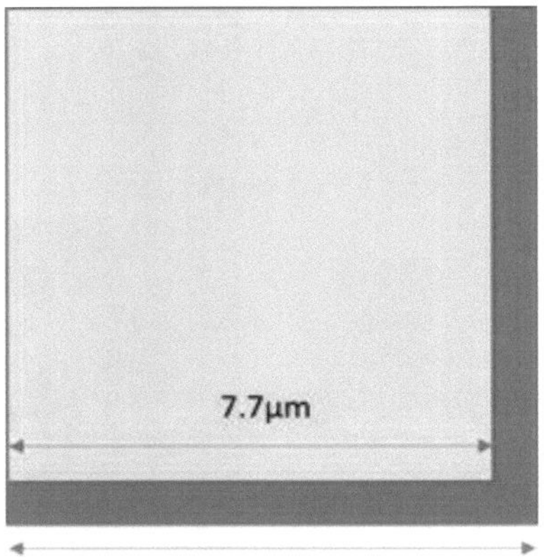

Figure 3.
Conceptual scheme of an individual pixel of a spatial light modulator for p = 8 μm and fill factor = 93%. Active pixel area is shown in light gray, interpixel gap is shown in dark gray.

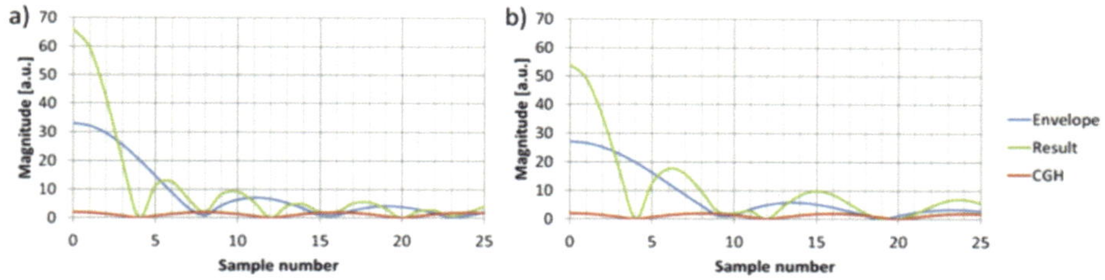

Figure 4.
Sinc envelope functions (blue) and resulting amplitude of holographic images for (a) pixel pitch equal to pixel size and (b) mismatch at 17%.

without a significant increase in the complexity or size of the optical setup. In larger experimental setups, for example, in the far-field holographic projection of 2-D intensity patterns, one can easily use filtering of the Fourier spectrum [6], as presented in **Figure 5**.

With this technique, one can efficiently filter out higher diffractive orders and additionally zero diffractive order, which is understood as the light that was not successfully modulated by the SLM. Depending on the size of the optical setup, the placement and transparency of the filter must be precisely matched to the beam; therefore, one of the proposed techniques is the photographic in-situ exposure and development of a perfectly matched filter [7], presented in **Figure 6**.

The advantage of this method is that it allows virtually all spatial frequencies achievable by the SLM to be used for image projection, except for the zero frequency. On the other hand, it requires a long optical path and access to the Fourier plane, which is usually problematic in compact devices.

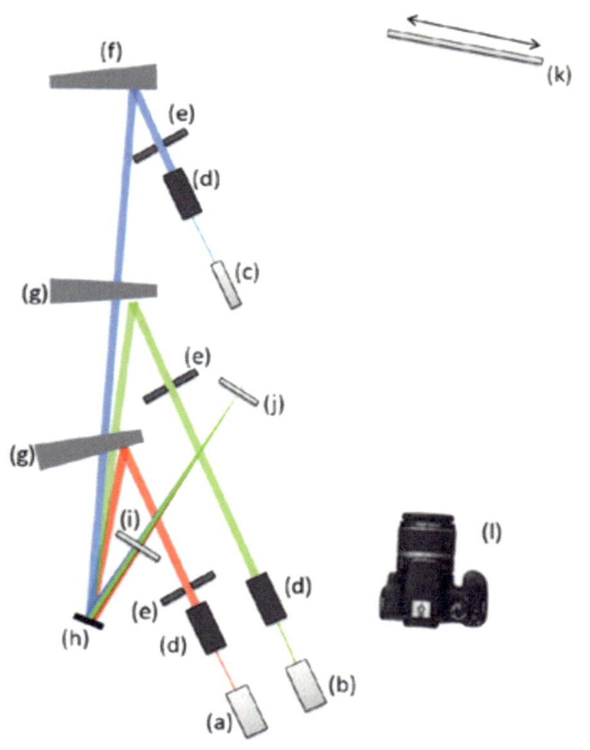

Figure 5.
Experimental setup of holographic projection with filtration in the Fourier domain [7].

Figure 6.
Matched Fourier-plane filter, exposed and developed in the optical path [7].

3. Illumination of the SLM in compact optical setups

In long-distance projection, one has the freedom of choosing the scheme of illumination of the SLM. Typically, one can use a beam splitter configuration, as shown in **Figure 1**, for the price of the loss of c.a. 75% of the light by two passes through the cube. If the energetic efficiency and elimination of stray light is the critical factor, the more suitable illumination scheme is the tilted mode presented in **Figure 5**, where the inclined beams are colinearly reflected from the SLM at angles derived from the equation of a diffractive grating. Obviously, for larger angles of incidence, the SLM will introduce astigmatism, as a result of a slightly denser effective pixel structure in one direction. This can be taken into account by multiplication of the displayed CGHs, for eample, with the phase pattern of a properly matched cylindrical lens or by applying phase correction by calculation based on Zernike coefficients. Nevertheless, the above mentioned simple techniques are of little importance in compact projection setups designed for near-eye displays and other portable, lightweight devices. The projection in such setups typically assumes the formation of a real image inside the device, which means projection distances of roughly few centimeters. The theoretical minimal projection (focusing) distance of an SLM with a pixel pitch of p and aperture of D is given by [5]:

$$z = \frac{D}{2\tan\left(\arcsin\frac{\lambda}{2p}\right)} \tag{2}$$

In order to decrease the aforementioned distance, one can increase the optical power of the SLM by adding a physical lens in the plane of the modulator, as shown in **Figure 7**. In addition, a more compact light source can be used in a form of a polarization-maintaining (PM) single-mode optical fiber with the coupled

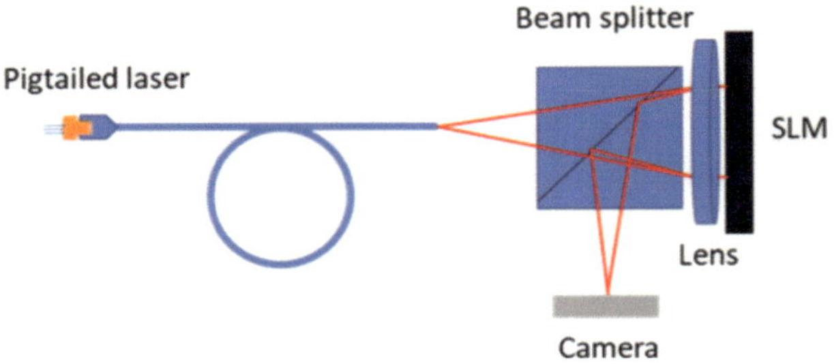

Figure 7.
Compact illumination scheme of the SLM with additional focusing lens.

laser diode. Due to the small diameter of the fiber's core, such a light source can be treated as quasi-point, yielding diffraction-limited imaging resolutions.

The advantage of this method is that as a result of almost zero distance between the SLM and the focusing lens, they can be treated as a single lens with easily adjustable focal power. As a result, the combined optical aberrations of such a tandem can be easily corrected with proper phase masks displayed on the SLM. Therefore, virtually aberration-less imaging can be realized in an extremely small configuration.

The use of a beam splitter allows the proper SLM incidence angle of zero degrees but wastes most of the energy of the light source to stray light. The below configuration shown in **Figure 8** helps to overcome this problem for the price of using custom-designed free-form optics [8].

The freeform prism is illuminated with a divergent beam from a bare laser diode equipped with a half plate for the control of polarization. The angles are matched so as the rays are reflected from the curved upper surface of the prism according to the rule of a total internal reflection (TIR). The reflected wave has lower divergence as if it has been transformed by a concave mirror. The beam that reaches the SLM has an incidence angle of almost zero degrees; therefore, the optimal conditions of phase modulation are met. The wave reflected from the SLM passes through the prism, again being focused at the exit to the air by the

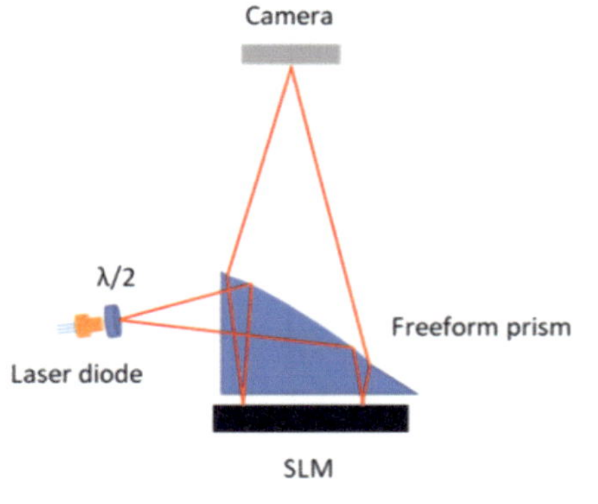

Figure 8.
Side illumination of the SLM with freeform prism and the utilization of the total internal reflection [8].

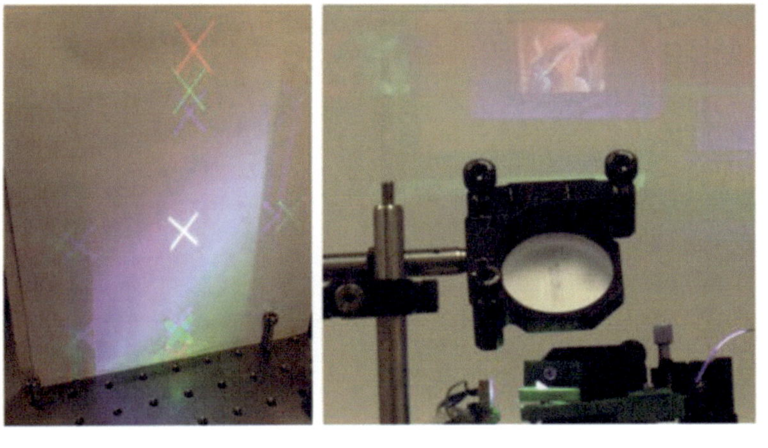

Figure 9.
Experimental realization of miniature projection head with side illumination allows efficient color projection in ambient lighting conditions [8].

curved interface of the prism. The exit beam has collimated nature, which can be easily changed to a convergent one by adding small optical power by the SLM. For this reason, the depicted setup is suitable for far-field long distance projection with extremely small and efficient optical setup, as seen in the practical realization shown in **Figure 9**.

4. Computation of computer-generated holograms

In the Fourier holographic projection configuration, shown in **Figures 1, 7** and **8**, the projected image is the intensity Fourier transform of the field rendered by the SLM. In practical situations when asymmetrical images are played back on the screen, the SLM would have to form a complex field, which is currently impossible, unless methods of complex modulation on double SLMs are used, see subchapter "Complex modulation method".

4.1 Random and ordered phase approach

For phase-only SLMs, the amplitude part of a complex field is discarded; therefore, the main function of the CGH computation algorithms is the transfer of object information from a complex field to the phase domain. Two of the most general and universal algorithms: Gerchberg-Saxton (GS) [45] and Random Phase Free [9]—assume that initially the phase domain of the hologram is filled with spatial frequencies that should form a quasi-uniform intensity at the projection plane. In the next step, the algorithms in an iterative loop let the phase evolve freely by cyclical applications of amplitude constraints in the image plane and in the SLM plane. The main difference between the mentioned algorithms is the character of the initial phase. It is random in the G-S algorithm and ordered in the RPF algorithm, nevertheless, both configurations simulate physically feasible situations, depicted in **Figure 10** [10].

In the random phase approach (G-S), the CGH on the SLM is supposed to contain all the spatial frequencies that allow each point of the image to be created

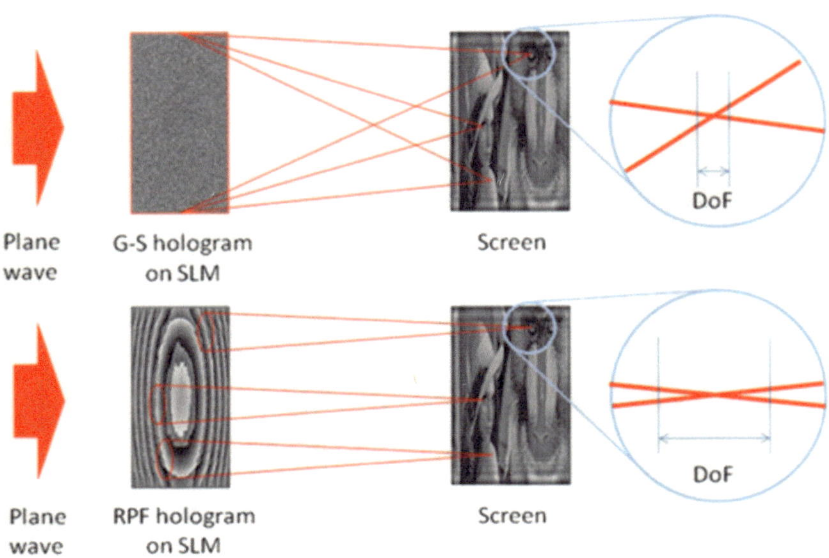

Figure 10.
Comparison of image forming by holograms calculated with G-S and RPF algorithms [10].

by the entire surface of the modulator. In other words, each portion of the SLM creates the entire image—see the upper half of **Figure 10**. In contrast, the ordered phase approach allows the minimal set of frequencies so that each point on the SLM creates only a small portion of the image—in other words, each point of the image is created by a small fraction of the SLM—as shown in the bottom half of **Figure 10**. The big advantage of the ordered phase approach is the smooth and slowly variable phase with sparse zones, as compared to G-S results. Such distributions are easier to be displayed efficiently by the SLM because the number of modulo-2π jumps is much lower and the small differences between the values of adjacent pixels of the SLM minimize the influence of cross-talk, hence increasing the overall diffractive efficiency.

4.2 Depth of focus in holographic projection

The interesting consequence of the varying set of spatial frequencies in CGHs is the different depth of focus of sharp projection. It is much larger in the RPF approach since that configuration can be considered a set of multiple imaging setups of very high F (aperture) numbers. The G-S configuration, on the other hand, can be understood as a single imaging setup with low F-number, and hence the lower depth of focus is natural here. **Figure 11** shows the experimental comparison of both algorithms reconstructing a static image of the *USAF-1951* test pattern at the projection distance of 1000 mm with different intentionally introduced defocus values. Note that the noise amount in RPF is much lower.

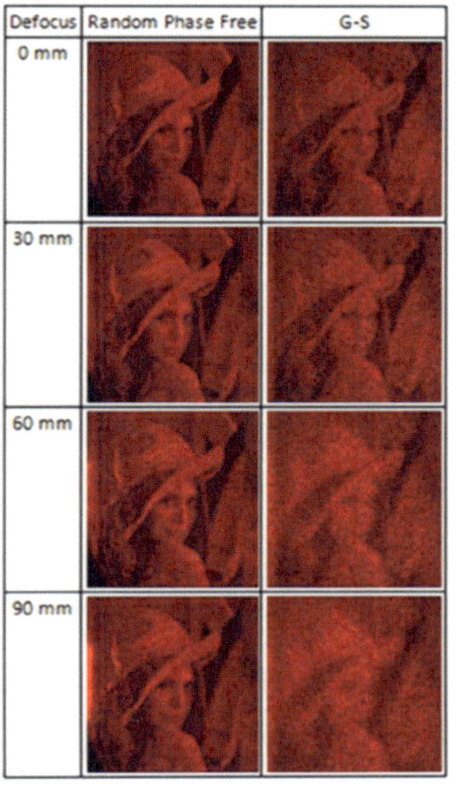

Figure 11.
Comparison of experimental projections from G-S algorithm and RPF algorithm under different defocus for base projection distance of 1000 mm [10].

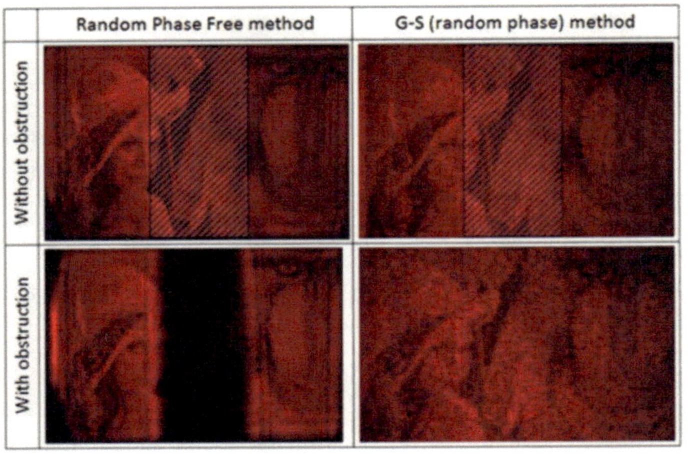

Figure 12.
Resistance to large obstruction of the SLM visible in experimental projections from CGHs calculated with RPF method and G-S method [10].

4.3 Resistance to local defects and obstructions of the SLM

The unique feature of holography is the robustness against any local defects and obstruction at the plane of the light modulator. This is the consequence of the fact that in the diffusive-type CGHs, each point of the image is formed by the whole SLM, therefore obstruction or even a large number of dead pixels is not imaged onto the projected image. This useful feature is especially important for potential applications in portable devices, which tend to wear out and get dirty or damaged. Moreover, from the practical point of view, the quality inspection of SLMs can be far less strict, and fully functional projectors might be constructed from faulty modulators from production rejects. This should lead to greater yield and lower prices of SLMs in the future.

Although the RPF method of CGH computation is superior in many areas, its resistance to local defects on the SLM is much lower than that of the G-S, as shown in **Figure 12**.

This subchapters presents the pros and cons of two different CGH computation algorithms, which belong to a quickly growing class of CGH computation methods [11, 12]. The proper choice of calculation method must meet the requirement and constraints of the planned application [13]. Both algorithms are reasonably simple, based on standard FFT (Fast Fourier Transform) and can be performed in real time on modern GPU (Graphics Processing Unit) processors [9, 14, 46].

5. Improving the quality of holographically projected images

5.1 Image resolution

The number of resolved points in the holographically projected images is close to the number of physical pixels of the SLM taking action in phase modulation, provided that the optical aberrations are carefully taken care of. Thanks to those experimental projections with full HD resolution have been successfully demonstrated [15]. On the other hand, a further increase of the informative capacity of the projected images would require larger arrays of pixels in SLMs with smaller and smaller pixel pitch. Such an approach is very demanding technologically, and the cross-talk between adjacent SLM pixels [16] would become the dominant effect, compromising the depth and speed of phase modulation. This miniaturization is not to be expected soon enough, therefore another concept of increasing the effective aperture of the SLM was proposed [17].

The Synthetic-Aperture SLM (SASLM) assumes the precise side-by-side positioning of two or more identical SLMs so that under common coherent illumination, they would form a single large SLM from the optical point of view. **Figure 13a** shows the experimental realization of SA-SLM.

The optical flatness of each SLM used in the coherent array is a strict requirement which allows the proper interference of fields coming from SLMs to form a common, high-resolution projected image. This can be measured, for example, with an interferometer and corrected by adding proper corrective phase to the contents on the SLM, as shown in **Figure 13b** and **c**. The use of two SLMs in tandem allows the precise guiding of two beams meeting in the acquisition plane, as seen in **Figure 14**.

The higher localization of a point-spread function spot (PSF) [18] in the Fourier plane of the setup was measured, as shown in **Figure 15**, potentially yielding two-fold increase of image resolution in the direction set by the orientation of the array of SLMs.

5.2 Coherent noise in projected images

Typically, holographically projected images are noisy and there can be named at least two origins of such coherent speckle-like noise. The first one is connected with the speckle noise from a diffuse screen on which the projection takes place. The local roughness of the screen causes uncontrolled interferences when the highly coherent projection beams are diffused toward the viewer's eyes. Numerous

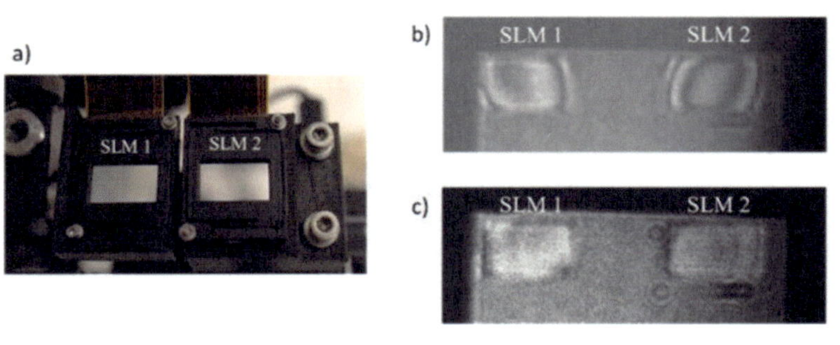

Figure 13.
Experimental setup showing the feasibility of a synthetic-aperture SLM in holographic image projection: (a) precisely positioned SLMs; (b) interferograms of SLMs before correction and (c) after correction [17].

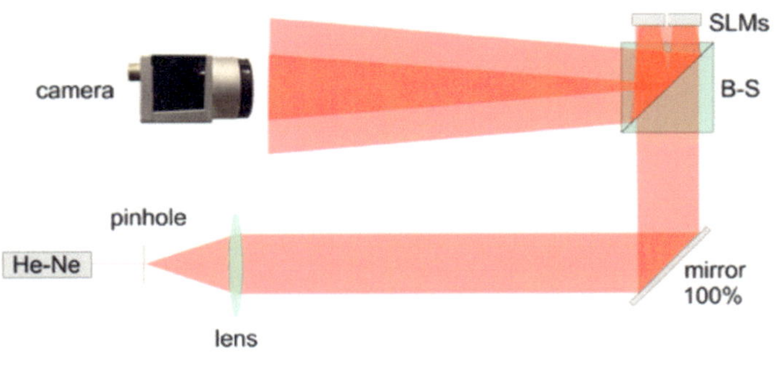

Figure 14.
Optical setup for the realization of improved light focusing by synthetic-aperture SLM [17].

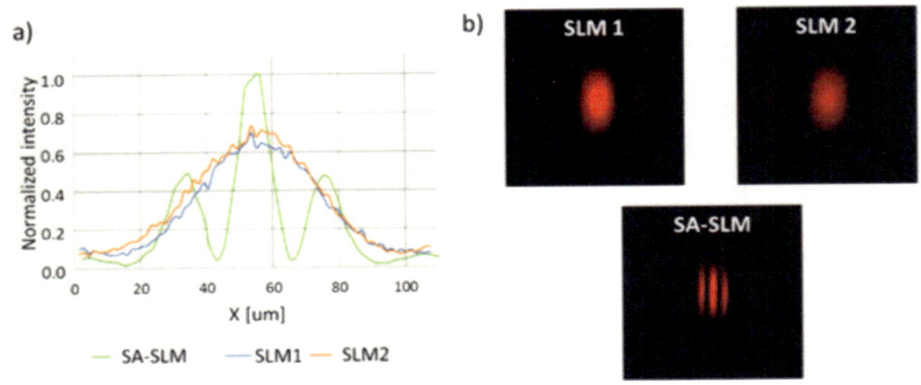

Figure 15.
Experimental measurements of the PSF spot shrunk in the x-direction as a result of synthetic-aperture SLM: (a) cross-sections; (b) captured intensity fields [17].

chapters can be found about the mitigation of such noise, some of them show the use of piezo-actuators to laterally move the screen or a selected optical element inside the projector. Apart from mechanical and polarization-based modulation not much can be done with this source of noise at the stage of design and computation of a CGH.

The other origin of the noise is contained in the algorithm of CGH computation. As a consequence of the phase-only modulation of the SLM, special algorithms of CHG computations must be used which yield unitary amplitude and decent projection quality from the holographic information stored only in the phase component [19, 20]. One of the most popular algorithms is the Gerchberg-Saxton method (see subchapter "Computation of Computer-Generated Holograms"), which uses random phase to create an initial distribution containing a large set of spatial frequencies that the SLM is capable of carrying. In the iterative loop, those frequencies are eliminated based on the constraints in the image plane (usually the image amplitude) and in the SLM plane (usually equal, unitary amplitude). The object information is lost in the process; therefore, the missing frequencies cause speckle noise in the reconstructed images. The physical mechanism behind this phenomenon is the lack of precision in setting the correct phase relations between light rays forming the adjacent image points. Obviously, increasing the number of iterations, one can improve the phase relations and decrease the noise amount, but the computation times then are impractical. Therefore, one must assume the presence of CGH-based noise in holographic projection. The next subchapters will discuss two proposed methods of suppression of this noise, which exploit the long integration time of a human eye and electronic detectors.

5.2.1 Time-domain noise averaging

Initial random phase distribution partly determines the position of phase ambiguity points in the intensity field reconstructed from a CGH. Therefore, the position of bright and dark grains of speckle noise can be altered by re-randomization of the initial phase in the Gerchberg-Saxton algorithm, while the signal information remains unchanged. Therefore, one can quickly display a sequence of holograms of the same object calculated with different initial phase distributions to obtain an effective and simple time-averaging of the noise, which can simulate the partly incoherent illumination [21]. If the speckle noise contrast is defined as:

$$C = \frac{\sigma_I}{<I>}, \tag{3}$$

then the noise averaging technique allows the decrease of C as $\sqrt{N}$, where N is the number of time-integrated frames. In practical experiments, one can decrease the typical noise ratio values from approx. 50% down to 10% by integrating 25 sub-holograms in a single exposure to be examined [22]. **Figure 16** shows exemplary experimental results of such noise integration, combined with piezo-movements of the projection screen.

5.2.2 Pixel separation method

The aforementioned uncontrolled phase relations in the light rays forming adjacent points in the image cause randomly occurring destructive and constructive interferences (see **Figure 17**), which greatly influence the final intensity pattern and thus increase the error [22, 23].

In order to avoid the uncontrolled interferences between closely packed image points, one can separate those points by splitting the input image into its sub-components having regular empty space between pixels, as shown in **Figure 18**. Those subcomponents are then the input images for regular G-S computations and finally are displayed on an SLM in a time sequence, giving a sequence of undisturbed reconstructions.

Obviously, if a fast frame rate SLM is used, the fragmentary images will integrate into the sensor giving the complete image with a very low amount of speckle noise.

The similar technique to time-domain pixel separation is the tiling of the elementary hologram on the surface of the SLM [24, 25] for the price of lowered image resolution.

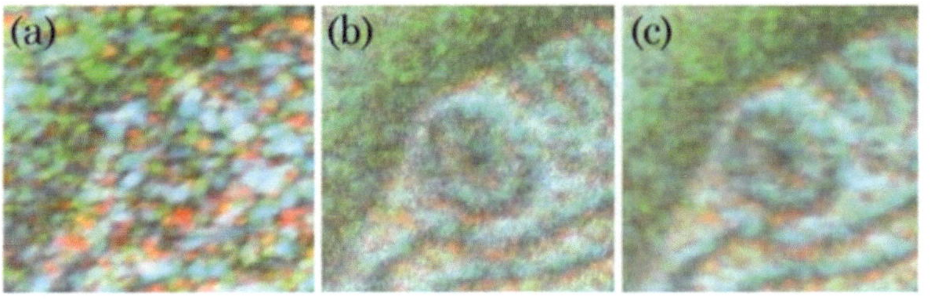

Figure 16.
Close-up of color reconstruction from a CGH computed with the Gerchberg-Saxton algorithm: (a) original image; (b) with time-integration of 25 sub-holograms; (c) with time-integration and piezoelectric movements of the projection screen [7].

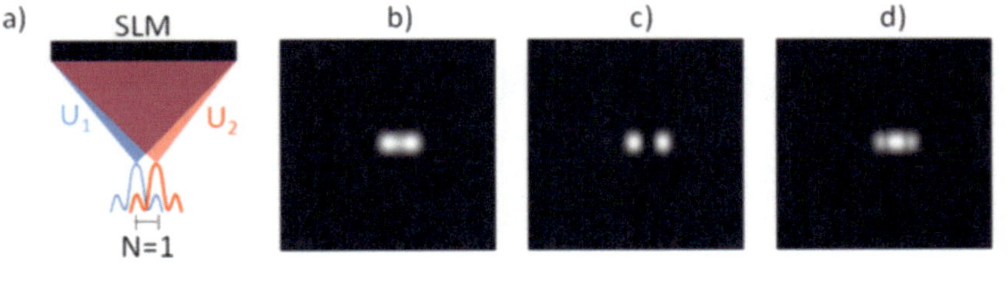

Figure 17.
(a) Interference between fields forming densely packed image points for (b) incoherent illumination; (c) coherent illumination with destructive interference of field between image points; (d) with constructive interference [22].

The comparison of the two presented methods of noise suppression is given in **Figure 19** [22], where the pixel separation methods allow a roughly threefold decrease of noise for the same number of integrations.

Both presented methods require very fast SLMs, therefore are optimal rather for electro-holography based on DLP/DMD [26, 27] or FLCoS (ferroelectric liquid crystal on silicon) modulators [28] with kHz refresh rates. On the other hand, binary phase or amplitude modulation increases the light losses in the holographic process, therefore again a trade-off between efficiency and image quality must be found.

5.2.3 Complex modulation method

The presence of coherent noise is mainly caused by the loss of information by constraining electro-holography to phase-only modulation of light. The attempt to restore complex modulation [6, 29, 30] (i.e., simultaneous amplitude and phase modulation) gives three important advantages in CGH: elimination of speckle noise, faster (i.e., non-iterative) computation and the freedom of phase distribution in the played-back fields. Currently, there are no commercially available complex modulators with appropriate pixel pitch and resolution, therefore one must use at least two separate SLMs and construct an optical setup which would

Figure 18.
Reconstructed subcomponents of a USAF-1951 test pattern created for the pixel separation method [22].

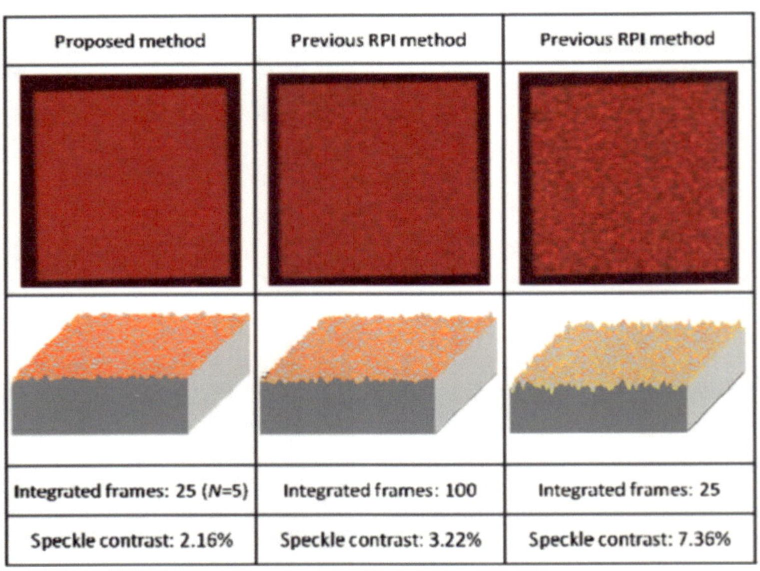

Figure 19.
Comparison of the close-up of the reconstructed uniform area from the USAF-1951 test pattern with the pixel separation method (denoted as "proposed method") and time averaging (RPI—random phase integration) for a different number of integrated sub-holograms [22].

compose them into a single complex modulating plane from the optical point of view. The conceptual scheme of this approach is shown in **Figure 20** [31]. The concept assumes the projection of the desired amplitude of the complex hologram with the use of first SLM on the surface of the other. Then (assuming the pixel-in-pixel positioning) the second SLM adds the proper phase delays of the complex hologram to the process. Then such created complex field propagates toward the projection screen or a detector camera.

It is important to position the SLMs with pixel precision and not to introduce any aberrations to the wave propagating from one SLM to another. For these reasons, the simple imaging of one SLM onto the surface of the other with 1:1 magnification (relay optics) usually introduces too much error and makes the experiment very difficult to adjust [32]. The lack of additional optical elements requires that the CGH computation algorithms take into account the nonzero distance between the SLMs, as seen in **Figure 20**. The numerical propagation at this distance is incorporated into the iterative optimization of the phase of such hologram yielding successful modulation resulting in virtually noiseless reconstructions, see **Figure 21**.

The aberrations caused by the intrinsic curvature of both SLMs must be thoroughly corrected; otherwise, the phase relations would dramatically change in the peripheral areas of the SLMs [32].

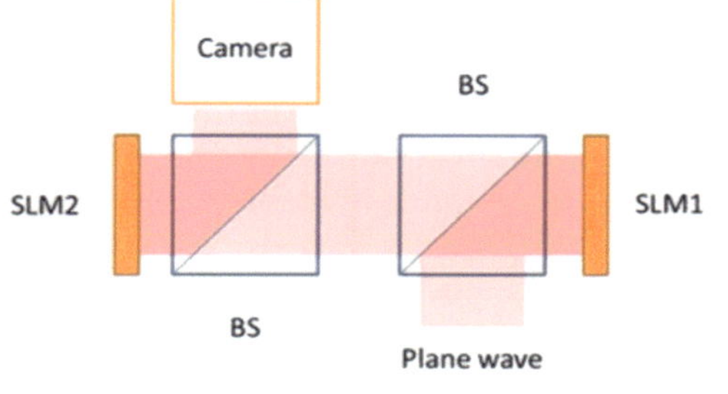

Figure 20.
Conceptual scheme of complex light modulation on two aligned SLMs with two beam-splitter (BS) cubes [31].

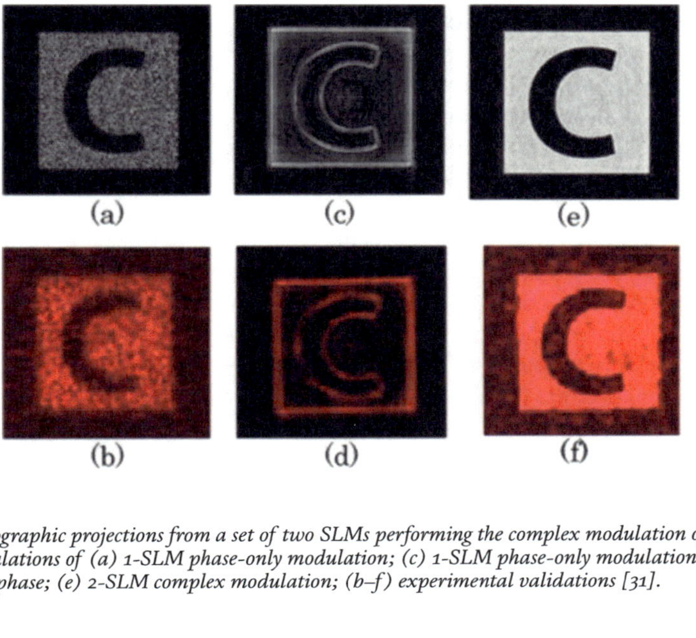

Figure 21.
Exemplary holographic projections from a set of two SLMs performing the complex modulation of light. Numerical simulations of (a) 1-SLM phase-only modulation; (c) 1-SLM phase-only modulation without random initial phase; (e) 2-SLM complex modulation; (b–f) experimental validations [31].

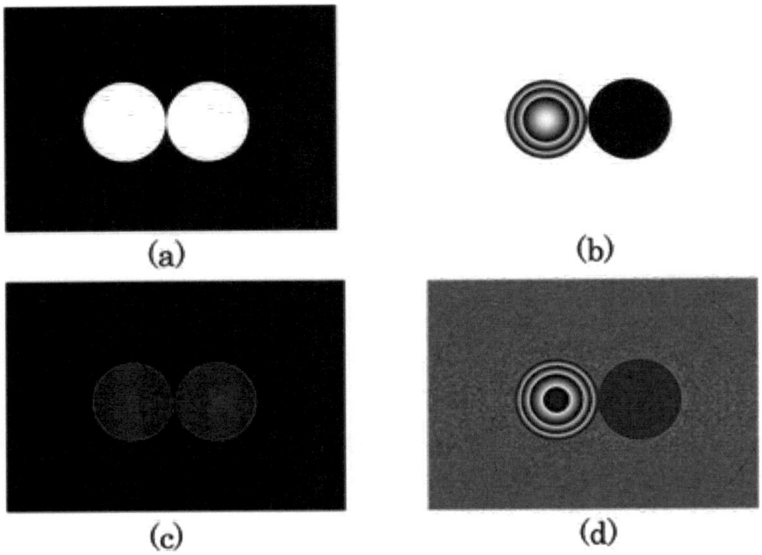

Figure 22.
Numerical simulations of the amplitude and phase projections using computer holography with complex light modulation: (a) desired amplitude; (b) desired phase; (c) numerically projected amplitude; (d) numerically projected phase [32].

The application of complex modulation gives the freedom of adjusting the phase state of the displayed field. Numerical simulations of such holographic reconstructions predict that one could control independently the amplitude and phase at the projection plane, as shown in **Figure 22**. This idea may give significant benefits, for example, in optical trapping and other applications where the spin-momentum of light is of importance.

The described methods of suppression of CGH-based noise should be always combined with the proper choice of the computation algorithm. For example, the RPF algorithm [9] allows virtually noiseless projection.

6. Holographic projection in color

The spatial light modulators are intrinsically monochromatic displays, and all the used rules of diffraction are a constraint to a specific, known wavelength. Without any additional precautions, the strong color mismatch occurs in the holographic projection of color images. As seen in **Figure 2**, the primary components of a projected color image are holographically formed at vastly different angles due to intrinsic chromaticity of the set of diffractive gratings displayed on the SLM. As a first approach, one could use three separate SLMs for the processing of three separate set of holograms, each computed for the given primary wavelength. Nevertheless, this solution greatly complicates the optical setup and implicates higher cost of any projection device. In response to that, this subchapter covers selected methods of displaying color contents with the use of electro-holography on a single SLM. The choice of the optimal method must be based on the characteristics of the SLM. If the frame rate is high the time-division method is favorable, as it allows the full image resolution and extreme simplicity of the illumination. If the SLM lacks fast response and provides a high pixel count, the spatial division method should be considered. The multi-plane method is favorable if high CGH computing power is available on site and the images to be displayed lack high spatial frequencies.

6.1 Compensation of the chromaticity of the SLM

In computer-generated holography, the SLM displays a complex set of diffractive gratings, which obviously diffract different wavelengths at variable angles. The easiest way of quick cancelation of this effect is the initial resize of the color components of the input color image so as to take into account the chromatic dispersion of the process holographic projection. **Figure 23** shows the properly resized and repositioned input images adapted for tilted-illumination projection with the use of three lasers of wavelengths: 633, 532 and 488 nm [33].

6.2 Color holographic projection by time division

The most intuitive and simple way of extending the SLM operation to color imaging is the time division. It assumes the illumination of the whole surface of the SLM in a sequence, for example, Red- > Green- > Blue- > Red- > ... . Obviously, the contents displayed on the SLM must be matched with the illuminating wavelength [34]. This requires quick refresh rates of the SLM because effectively the native frame rate of the panel is divided by three. Practically, the minimal native refresh rate of the SLM is 180 Hz, which allows 60 frames per second of the final color holographic animation. This, in turn, requires the use of especially designed liquid crystal cells [35] or switching to binary ferroelectric LCs [36]. **Figure 24** shows an exemplary experimental holographic projection of color images with the use of time division for pixel separation and color separation.

6.3 Color holographic projection by spatial division

In order to overcome the problem of color breakup and use the full achievable frame rate of the SLM, the spatial division [37, 38] may be used instead of time-division. It involves dividing the SLM into three separated regions illuminated by three beams with associated colors and known wavelengths [39]. **Figure 25** shows the extreme simplicity of this approach [40].

The high-quality color filters are applied directly to the surface of the SLM in order to avoid the cross-talk between regions [41]. Under the filters, three CGHs are displayed side by side, computed with the assumed wavelength, positioned at the projection screen by numerical complex multiplication with proper diffractive gratings and spherical lenses.

The technique is advantageous in a way that computing three smaller holograms is easy to program in parallel graphics processing units (GPU) and field programmable gate array (FPGA) systems [42, 43], and in that, the full frame rate of the SLM is used with no color breakup and smooth color display. This occurs for the inevitable price of lower image resolution due to a limited number of pixels devoted

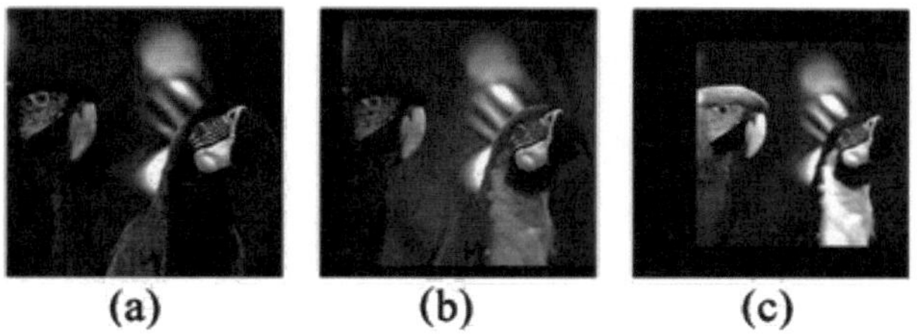

Figure 23.
Resizing and repositioning of the input color components allows quick compensation of the chromaticity of holographic projection: (a) blue component; (b) green component; (c) red component [33].

to the holographic reconstruction of a color component. **Figure 26** shows some exemplary colorful holographic projections [40].

6.4 Multi-plane color hologram

This method combines the advantages of the aforementioned techniques in a way that it utilizes the whole surface and resolution of the SLM in the given moment of time and additionally all three colors are displayed simultaneously [44]. On the other hand, it requires iterative holographic computations and is rather suitable for real-life images with dominant low spatial frequencies.

The chromaticity of SLM is typically a problem, while in this technique, it is treated as a useful phenomenon. First, a CGH containing three amplitude objects located at three distant planes is calculated using the iterative ping-pong algorithm originating from the Gerchberg-Saxton algorithm [45]. When such hologram is

Figure 24.
Exemplary color holographic projections with time division method: (a) input images; (b) captured images; (c) close-ups of the central region showing low noise, high resolution, and good color rendering [22].

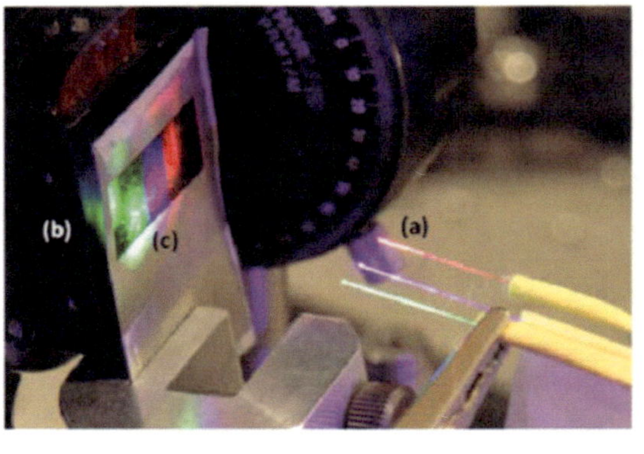

Figure 25.
Spatial division of the SLM: simple illumination with three fibers and color filters: (a) optical fibers; (b) SLM; (c) color filters [40].

DOI: http://dx.doi.org/10.5772/intechopen.85118

reconstructed with a single laser beam of a given wavelength (e.g., red), the three images are played back at projected distances. When a second laser beam of a different wavelength (e.g., green) simultaneously illuminates the SLM, the same images are played back in a different set of distances, according to the chromatic characteristics of the gratings displayed on the SLM. The same happens with the third beam (e.g., blue). One can use the degree of freedom in the choice of propagation

Figure 26.
Exemplary colorful holographic projections achieved with the spatial division method [40].

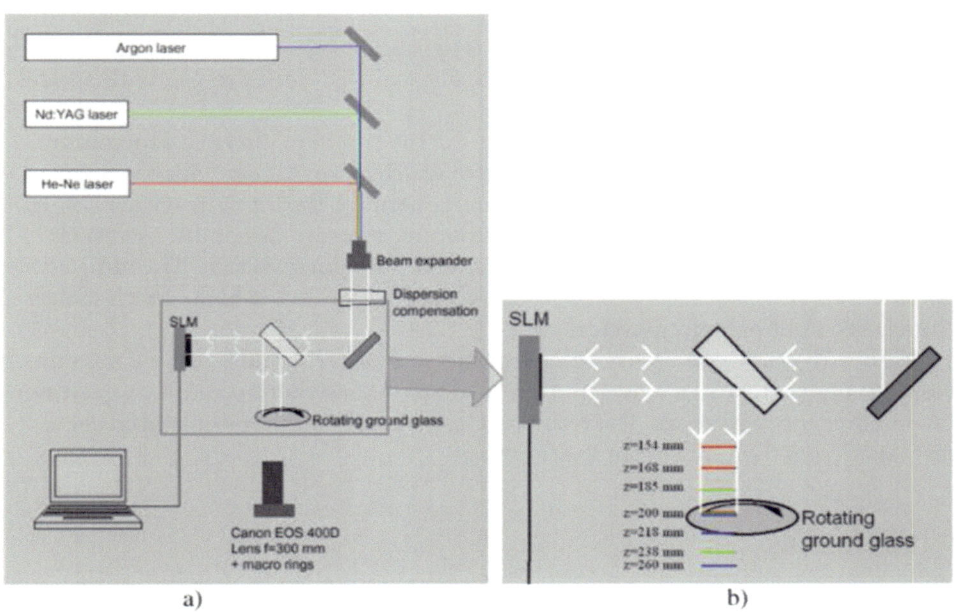

Figure 27.
Reconstruction of the color image at z = 200 mm with simultaneous illumination of the SLM with three wavelengths and displayed multi-plane CGH [44].

Figure 28.
Exemplary color reconstructions from multi-plane computer-generated hologram [44].

distances of three encoded image planes in order to allow the reconstruction of three images at a fixed plane with three projected wavelengths. This process is depicted in **Figure 27** [44].

The important feature of this method is the simplicity of SLM illumination with three colinear laser beams, which can be easily achieved, for example, with optical fibers [46].

The quality of the color image perceived at the common distance depends on the convergence of the iterative algorithm of CGH computation, which is a constraint for the situation where the color components of the input image are similar. Otherwise, the field creating three very different intensity distributions at such closely packed planes becomes nonphysical and the algorithm fails. The same occurs when the spatial frequencies in any of the three images are too high, which makes the desired light field nonphysical.

Figure 28 shows exemplary color reconstructions from multi-plane holograms. Note the high agreement between the numerical reconstruction and the experimental outcome for most cases. The results are compared with time-sequential color display (denoted as *3 SLMs*) in the figure.

7. Conclusions

Spatial light modulators based on liquid crystals gained popularity in research groups dealing with real-time computer-generated holography. The main reason was the well-established technology of micro-displays known from projectors, monitors, and large screen TVs. On the other hand, their further miniaturization is problematic because one cannot simultaneously shrink the LC particles, and therefore the smaller and smaller pixel pitch inevitably leads to numerous technical problems like cross-talk, long response, poor fill factor, and flicker. Great effort in many research groups in the world was put to tweak the SLMs and obtain the best possible holographic quality from them, but in the future research of holographic TVs [47] and near-eye displays, their intrinsic hardware drawbacks (like fixed Cartesian pixel array) would be the severe bottleneck.

Next generation of SLMs should be "pixel-less", that is, have freedom of placement and shape [48] of light modulating microareas, and the used medium should not put constraints on the minimal size and dense packing of such "pixels." The problems that remain to be solved is to find the proper medium [49, 50] and the fast and precise method of its addressing, writing, and erasing. Until then, this practical guide to experimental use of LCoS SLMs in computer holography may be used to overcome their current hardware deficiencies.

Acknowledgements

Selected parts of this work were supported by the TEAM-TECH programme of the Foundation for Polish Science co-financed by the European Union under the European Regional Development Fund ("HANEDA", TEAM TECH/2016-3/18, POIR.04.04.00-00-3DD9/16-00).

Conflict of interest

The author declares no conflict of interest.

Author details

Michał Makowski
Warsaw University of Technology, Warsaw, Poland

*Address all correspondence to: michal.makowski@pw.edu.pl

References

[1] Hong K, Oh K, Choo H, Lim Y, Park M. Viewing window position control on holographic projection system by electrically focused tunable lens. Practical Holography XXXII: Displays, Materials, and Applications. 2018;105580R

[2] Wakunami K, Hsieh P, Oi R, Senoh T, Sasaki H, Ichihashi Y, et al. Projection-type see-through holographic three-dimensional display. Nature Communications. 2016;**7**(1):12954

[3] Lizana A, Lobato L, Marquez A, Iemmi C, Moreno I, Campos J, et al. Study of liquid crystal on silicon displays for their application in digital holography. Advanced Holography—Metrology and Imaging. 2011

[4] Cirino GA, Verdonck P, Mansano RD, Pizolato JC Jr, Mazulquim DB, Neto LG. Digital holography: Computer-generated holograms and diffractive optics in scalar diffraction domain. Holography—Different Fields of Application. InTechOpen; 2011

[5] Haist T, Osten W. Holography using pixelated spatial light modulators—Part 1: Theory and basic considerations. Journal of Micro/Nanolithography, MEMS, and MOEMS. 2015;**14**(4):041310

[6] Qi Y, Chang C, Xia J. Speckleless holographic display by complex modulation based on double-phase method. Optics Express. 2016;**24**(26):30368

[7] Makowski M, Ducin I, Kakarenko K, Kolodziejczyk A, Siemion A, Siemion A, et al. Efficient image projection by Fourier electroholography. Optics Letters. 2011;**36**(16):3018

[8] Makowski M, Kowalczyk A, Bieda M, Suszek J, Ducin I, Shimobaba T, et al. Miniature holographic projector with cloud computing capability. Applied Optics. 2019;**58**(5):A156

[9] Shimobaba T, Ito T. Random phase-free computer-generated hologram. Optics Express. 2015;**23**(7):9549

[10] Makowski M, Shimobaba T, Ito T. Increased depth of focus in random-phase-free holographic projection. Chinese Optics Letters. 2016;**14**(12):120901-120905

[11] Mengu D, Ulusoy E, Urey H. Non-iterative phase hologram computation for low speckle holographic image projection. Optics Express. 2016;**24**(5):4462

[12] Duan J, Liu J, Hao B, Zhao T, Gao Q, Duan X. Formulas of partially spatial coherent light and design algorithm for computer-generated holograms. Optics Express. 2018;**26**(17):22284

[13] Chang C, Wu J, Qi Y, Yuan C, Nie S, Xia J. Simple calculation of a computer-generated hologram for lensless holographic 3D projection using a nonuniform sampled wavefront recording plane. Applied Optics. 2016;**55**(28):7988

[14] Shiraki A, Takada N, Niwa M, Ichihashi Y, Shimobaba T, Masuda N, et al. Simplified electroholographic color reconstruction system using graphics processing unit and liquid crystal display projector. Optics Express. 2009;**17**(18):16038

[15] Makowski M, Ducin I, Kakarenko K, Suszek J, Kowalczyk A. Performance of the 4k phase-only spatial light modulator in image projection by computer-generated holography. Photonics Letters of Poland. 2016;**8**(1):26-28

[16] Persson M, Engström D, Goksör M. Reducing the effect of pixel crosstalk

in phase only spatial light modulators. Optics Express. 2012;**20**(20):22334

[17] Kowalczyk A, Makowski M, Ducin I, Sypek M, Kolodziejczyk A. Collective matrix of spatial light modulators for increased resolution in holographic image projection. Optics Express. 2018;**26**(13):17158

[18] Běhal J, Bouchal Z. Optimizing three-dimensional point spread function in lensless holographic microscopy. Optics Express. 2017;**25**(23):29026

[19] Chang C, Qi Y, Wu J, Xia J, Nie S. Speckle reduced lensless holographic projection from phase-only computer-generated hologram. Optics Express. 2017;**25**(6):6568

[20] Utsugi T, Yamaguchi M. Speckle-suppression in hologram calculation using ray-sampling plane. Optics Express. 2014;**22**(14):17193

[21] Czerwiński A, Kakarenko K, Sypek M, Makowski M, Ducin I, Suszek J, et al. Modeling of the optical system illuminated by quasi-monochromatic spatially incoherent light: New numerical approach. Optics Letters. 2012;**37**(22):4723

[22] Makowski M. Minimized speckle noise in lens-less holographic projection by pixel separation. Optics Express. 2013;**21**(24):29205

[23] Mori Y, Fukuoka T, Nomura T. Speckle reduction in holographic projection by random pixel separation with time multiplexing. Applied Optics. 2014;**53**(35):8182

[24] Georgiou A, Christmas J, Moore J, Jeziorska-Chapman A, Davey A, Collings N, et al. Liquid crystal over silicon device characteristics for holographic projection of high-definition television images. Applied Optics. 2008;**47**(26):4793

[25] Makowski M, Kakarenko K, Ducin I, Kowalczyk A, Bieda M, Suszek J. Study of image resolution in holographic projection. Photonics Letters of Poland. 2014;**6**(3):96-98

[26] Su P, He Z, Ma J, Cao L, Yuan R. Design of color LED holographic display system based on DMD. Digital Holography and Three-Dimensional Imaging. 2016:W2A-16

[27] Park M, Lee B, Son J, Chernyshov O. Properties of DMDs for holographic displays. Journal of Modern Optics. 2015;**62**(19):1600-1607

[28] Schmieder F, Klapper S, Koukourakis N, Busskamp V, Czarske J. Optogenetic stimulation of human neural networks using fast ferroelectric spatial light modulator—Based holographic illumination. Applied Sciences. 2018;**8**(7):1180

[29] Cao L, Kong D, Zong S, Zhang H, Jin G, Xueju S. Complex wavefront modulation and holographic display using single spatial light modulator. Optics and Photonics for Information Processing XI. 2017:1039508

[30] Luis Martínez Fuentes J, Moreno I. Random technique to encode complex valued holograms with on axis reconstruction onto phase-only displays. Optics Express. 2018;**26**(5):5875

[31] Siemion A, Sypek M, Suszek J, Makowski M, Siemion A, Kolodziejczyk A, et al. Diffuserless holographic projection working on twin spatial light modulators. Optics Letters. 2012;**37**(24):5064

[32] Makowski M, Siemion A, Ducin I, Kakarenko K, Sypek M, Siemion AM, et al. Complex light modulation for lensless image projection. Chinese Optics Letters. 2011;**9**(12):120008-120010

[33] Makowski M, Ducin I, Sypek M, Siemion A, Siemion A, Suszek J,

et al. Color image projection based on Fourier holograms. Optics Letters. 2010;**35**(8):1227

[34] Han Z, Yan B, Qi Y, Wang Y, Wang Y. Color holographic display using single chip LCOS. Applied Optics. 2018;**58**(1):69

[35] Holoeye LETO SLM. Available from: https://holoeye.com/spatial-light-modulators/leto-phase-only-spatial-light-modulator/ [Accessed: January 20, 2019]

[36] ForthDD Products. Available from: https://www.forthdd.com/products/spatial-light-modulators/ [Accessed: January 20, 201901-20]

[37] Zaperty W, Kozacki T, Gierwiało R, Kujawińska M. RGB imaging volumes alignment method for color holographic displays. Photonics Applications in Astronomy, Communications, Industry, and High-Energy Physics Experiments. 2016:1003117

[38] Zaperty W, Kozacki T, Kujawinska M. Multi-SLM color holographic 3D display based on RGB spatial filter. Journal of Display Technology. 2016;**12**(12):1724-1731

[39] Shimobaba T, Takahashi T, Masuda N, Ito T. Numerical study of color holographic projection using space-division method. Optics Express. 2011;**19**(11):10287

[40] Makowski M, Ducin I, Kakarenko K, Suszek J, Sypek M, Kolodziejczyk A. Simple holographic projection in color. Optics Express. 2012;**20**(22):25130

[41] Tsuchiyama Y, Matsushima K. Full-color large-scaled computer-generated holograms using RGB color filters. Optics Express. 2017;**25**(3):2016

[42] Yamamoto Y, Nakayama H, Takada N, Nishitsuji T, Sugie T, Kakue T, et al. Large-scale electroholography by HORN-8 from a point-cloud model with 400,000 points. Optics Express. 2018;**26**(26):34259

[43] Shimobaba T, Ito T. Computer Holography: Acceleration Algorithms and Hardware Implementations. 1st ed. Boca Ranton, USA: CRC Press; 2019

[44] Makowski M, Sypek M, Ducin I, Fajst A, Siemion A, Suszek J, et al. Experimental evaluation of a full-color compact lensless holographic display. Optics Express. 2009;**17**(23):20840

[45] Gerchberg RW, Saxton WO. A practical algorithm for the determination of phase from image and diffraction plane pictures. Optik (Jena). 1972;**35**:237-246

[46] Kowalczyk A, Bieda M, Makowski M, Sypek M, Kolodziejczyk A. Fiber-based real-time color digital in-line holography. Applied Optics. 2013;**52**(19):4743

[47] Reichelt S, Haussler R, Leister N, Futterer G, Stolle H, Schwerdtner A. Holographic 3-D displays—Electro-holography within the grasp of commercialization. Advances in Lasers and Electro Optics. InTech; 2010

[48] Xiao Ma X, Juan Liu J, Zhao Zhang Z, Xin Li X, Jia Jia J, Bin Hu B, et al. Analysis of optical characteristics of modulation devices with square and circle pixels for 3D holographic display. Chinese Optics Letters. 2015;**13**(1):010901-010905

[49] Kveton M, Fiala P, Havranek A. Polymer holography in acrylamide-based recording material. Holography, Research and Technologies. IntechOpen; 2011

[50] Stupakiewicz A, Szerenos K, Afanasiev D, Kirilyuk A, Kimel A. Ultrafast nonthermal photo-magnetic recording in a transparent medium. Nature. 2017;**542**(7639):71-74

Chapter 6

Professor Avatar Holographic Telepresence Model

Luis Luevano, Eduardo Lopez de Lara and Hector Quintero

Abstract

Introduced into theaters in the 1860s, Pepper's Ghost startled theatergoers with an effect that allowed live people or objects to materialize into the scene. The illusion of a ghost is an actor located forward of and below the stage floor. The glass illustrates the reflection of the offstage "ghost," while the leftmost "ghost" simulates what the audiences see. Modern versions of this effect consist of a completely new way of projecting video to create the illusion of life-size, full-color, moving images but projected as 2D images into a set. The mind of the audience creates the 3D illusion. This technology enables a new line of communication, which is called "holographic telepresence" that delivers a life-sized holographic experience in real time, enabling to connect more effectively and make an impact on audiences. The technology reduces expenses and saves on time travel. This project identified the parameters for correct setup of holographic telepresence, so that future users will be able to replicate and use with ease. The project used action research, which provides fast and effective solutions. The results demonstrated that it was possible that by defining the parameters and a guide for setting up a holographic telepresence.

Keywords: hologram, telepresence, remote communication, holographic telepresence, holographic projection

1. Introduction

Communication over time has been evolving and improving so that people could have a much simpler and easier access toward the information that they need. Even before the emergence of technology, communication has been at the forefront of relationship building and business development. Ever since smoke signals, carrier pigeons, telegraph to the modern computer, and smartphones, the thing that is common among these technologies is the way it has changed how the human being communicates.

Newer advancements like texting and messaging apps have spurred even more efficiency within workplace communication. We have come a long way since the days of written letters and memos. Even email has become a secondary form of communication in the workplace as chat platforms are taking over. Advancements in communication continue to stimulate efficiency in every workplace. These advancements continue to improve and even to displace other efforts such is the case of written letters to email, which has become a secondary form of communication. This is where the concept of "telepresence" enters, and it consists of a combination of technologies that seek to represent a person that is in a distant location as if it was there. For this, it is necessary that the user can use his senses and obtain stimuli from the remote place. It communicates the position, movements, actions, and voice, and in some cases, there can be interaction with documents and other objects. So the information that is being interchanged is wide and rich in terms of type or kind of medium. Although in telepresence systems, it is important for the images to have a real-world scale. In telepresence systems, the goal is to make the user lose the notion that has some intermediate devices with a lot of technology and that can act in a natural way.

Over 30 years ago, MIT professor and artificial intelligence pioneer Marvin Minsky laid out an ambitious plan calling for the development of advanced tele-operated robotic systems that would usher in a "remote-controlled economy." He wrote about it in the science and science fiction magazine Omni in 1980. In his essay, Dr. Minsky envisioned a "remote-controlled economy." He coined the term "telepresence" to describe these systems, which in his futuristic vision would transform work, manufacturing, energy production, medicine, and many other facets of modern life [1]. He considered that the biggest challenge in developing telepresence is to achieve that sense of "being there." According to the company, Digital Video Enterprises, telepresence refers to technologies that allow a user to appear to be present, feel like they are present, or have some effect in a space the person does not physically inhabit. Telepresence can include video teleconferencing tools, where a picture and audio stream is conveyed to a remote location. It is a multidisciplinary combination of communication that integrates engineering, psychology, and the television broadcasting [2].

2. Pepper's ghost effect and its evolution

Since 1863, John Pepper and Henry Dircks developed the "Peppers Ghost Effect," by showing a play of Charles Dickens's The Haunted Man that basically involves a stage that is specially arranged into two rooms, one that people can see into or the stage as a whole and a second that is hidden to the side, the "blue room" (**Figure 1**). A plastic foil is a polymeric mirror with a specific formulation, thickness, and oriented in a way to maximize holographic effect. This film is angled so that whatever it will reflect can be hidden from the audience in a secret room. The hidden room is an entirely black mirror-image of the stage where the actual "ghosts" are placed. When it is time to make the ghosts appear in front of the audience, the hidden figures are lit and their reflection appears in the glass. The figures in the mirror-image room will be arranged so that their reflection corresponds with where they should appear on stage. For example, if you wanted to make a ghost appear at a table, the room visible to the audience would already have a table and a chair in it. However, in

Figure 1.
Configuration and projection of the Pepper's ghost effect on stage.

the mirror-image room, a figure would be sitting on a black chair, or similar prop, positioned so that the reflection lines up with the table and chair in the main room. When a light is turned on the figure, it will appear as though a ghost is sitting at the table visible to the audience [3].

Holograms have always been a subject of fascination by humans because it always seems as technology that comes from the future. Although there are already 3D holograms that give a tactile sensation, until now there is no technology that can produce a full body holographic projection of a person in any place. Currently, there is a new communication technology in development that will allow people to interact inside a control-simulated environment, even if they are thousands of kilometers apart [2].

Holography can create an accurate visual simulation, with total parallax: a replica of the real object made of light, which has the real object's visual properties but is immaterial, intangible. Holographic images appear to be three dimensional and with volume and depth which can be seen with the naked eye [4].

Since the early 1980s, there have been experiments in projecting holographic dynamic images and transmitting them remotely. The usual cinema and television images are built on the viewpoint decided by the filmmaker: the scene is created and presented through the filmmaker's eyes and perspective. The few spectators of the first 47-second monochromatic holographic movie, made in 1976 by the Russian scientist Victor Kumar, reported that they could see a young woman holding a bouquet of flowers [4].

3. What is holographic telepresence?

"It is a system that projects full-motion, realistic, and 3D images in real-time. A holographic telepresence system captures images of real, remote people and/or surrounding objects and compresses and transmits the images and sound over a broadband network." Once transmitted, it decompresses the images and finally projects them. It also includes real-time audio communication that further enhances the realism experience. In some cases, it could truly rival with the physical presence of a user [5].

Holographic telepresence is the next step of communication that consist a full-motion, 3D video conferencing system that can project distant people and objects in a room, with live feed audio and video communication, ranging from remote participation in meetings and conferences to virtual on-stage appearances at concerts [6].

In other words, "holographic telepresence it's the combination of one or more telepresence technologies with a holographic projection as the main medium of communication between users" [7]. A holoprojector will use holographic technology to project large-scale, high-resolution images onto a variety of different surfaces, at different focal distances, from a relatively small-scale projection device [8]. There are multiple intents and approaches in achieving a realistic holographic projection. Developers of the holographic display are working on a technology that will be used for teleconferencing. Thanks to its relatively low Internet bandwidth and computer processing requirements, a conversation between users that are being projected and transmitted would need the same bandwidth of a modern 2D video call [9].

4. Defining the problem

Standard video communication systems can be difficult to set up, challenging to use, and frequently unsatisfying in quality. Globalization has increased the need for remote telepresence systems that allow for remote collaboration among geographically dispersed colleagues and partners. Increasingly, business discussions must include not just multiple people and multiple work teams but multiple locations. Many of today's telepresence, video communication, and collaboration tools provide an enormous productivity boost; that is why they are being put as the next medium of communication, because what is sought is to break the time and distance barrier.

While these video conferencing systems can be conducted at any time of the day, replace many in-person business trips, increase productivity by eliminating many barriers by helping decisions to be made faster, they still lack personal interaction, and in some cases, meetings require a personal touch to be successful. Video conferencing can be less personal than meeting face to face, and it can be possible to miss vital body language when you are struggling with a pixelated image or stuttering video. Setting these kinds of video conferencing in an office can be a bit expensive for small-sized companies. Simple features can fit into the budget, but if advanced features are required, then a substantial amount of expenditure must be done.

Increasing the capability of a team of people, such as approaching a complex situation, gaining comprehension, and finding solutions; wherever they are in the world.

4.1 Justification

This research seeks to define a setup and identify and set up the parameters for remote holographic telepresence communication, and it is justified by flashy, pricey room systems which have been one of the most common deployments of videoconferencing technology in the workplace. These kinds of setups are still a facet of the executive conference room that use the latest video and audio systems that the market has to offer which leads to a high-end budget [18].

Achieving the humanization of the virtual remote contact, stimulating teamwork and academic collaboration will fundamentally change the form in which people will communicate in the future. Being a communication system, it is logical that the whole process will be direct, obtaining an interaction and communication channel to be as natural as face-to-face communication holograms [10]. Holographic telepresence combines technologies that already exist with a special care in the environment in which it takes place. The position of the cameras is fundamental so that the people that are receiving the transmission can appreciate real-world proportions and give continuity where it is being projected.

Holographic telepresence can revolutionize the way we think about and experience modern communication systems. In fact, it has the potential to change diverse types of communication systems. This type of technology can reduce the time, money, and effort otherwise wasted for traveling for business meetings or people that give conference or lectures. It may facilitate distance education like never before by connecting geographically remote classrooms, illustrating learning processes, and homogenizing the education level of schools and professors [11].

So why it is not more popular?

According to Arjona [12], most of the technical problems that impede a greater profusion of videoconference or remote telepresence technologies have been fixed long ago:

- Image quality: 2 Mbps is more than sufficient for obtaining a similar quality as of a normal television set. The new standard H.265 is still in a development stage, and it will support ultra HD images as far as 8 Mega pixels.

- Latency: the new processors are more capable of supporting the demanding requirements that video codecs require, which minimizes the time between capture, coding, transmission, and decoding of images making possible a more fluid line of communication.

- Flexibility: It is now possible to establish a videoconference via the Internet, perform video calls from almost any mobile device.

- Cost: the cost of webcams, laptops, projectors, and mobiles has decreased considerably.

Users are still showing signs of resistance to the use of these systems. Some of the main motives are:

- Speed connection: remote telepresence requires establishing different channels of communication for audio, video, and data, and in some cases, it requires a dedicated bandwidth; all of these require a vast amount of time.

- Localization: the user needs to be sited in a special room that is not always at their disposal.

- Naturalness: the camera is usually located on top of a screen or in a monitor, and when the user speaks to another person, they are looking in any other place but not in the other user's line of sight.

5. Defining the problem

The purpose of this work is to propose a setup and a method that can potentially improve and help so that others may easily carry out their own holographic telepresence communications with ease. As shown in **Figure 2**, this method will present the knowledge, basic requirements, and step-by-step guide, so that the users can understand and may save time in the installation and setup. This method is not intended to replace other forms of communication, but on the contrary, it is a proposal for enhancing the remote telepresence experience, so that its use can be further divulged and be as natural as other mediums of communication as a telephone, instant messaging, social networks, etc.

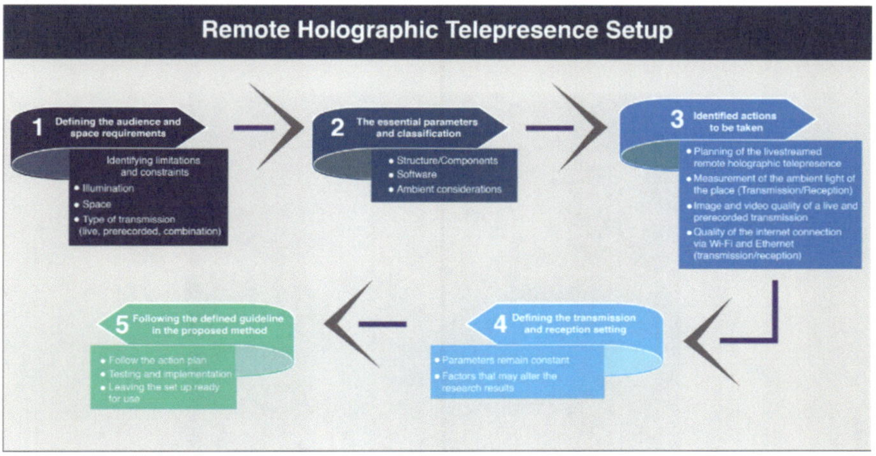

Figure 2.
Proposed solution for improving and accelerating the setting process of a remote holographic telepresence communication system.

5.1 Objectives

The motivation of this research is to improve the way holographic transmissions are taking place so that new users will be drawn to it for being practical and easy to use. Because sometimes setting these kinds of systems can be demanding, time-consuming and can demotivate and turn away potential users (**Figures 3–5**).

The main objective is to establish the parameters for the correct setting of a remote holographic telepresence system as a medium of communication and support. With the purpose of demonstrating, it is possible to improve and humanize long-distance communication and interaction. During the investigation, it is expected to accelerate the setting process of a holographic projection system for future presentations, by taking measurements of the distance of the projector depending on its luminosity, room illumination, room space, dress code, Internet bandwidth, etc.

This technology will increase the reach, impact, and remembrance between work colleagues and students. It will be the next evolutionary step of videoconference systems that are currently limited to a television screen, changing the way we communicate and travel around the world.

During the development of the project, the questions that we are looking to respond to are:

What are the constant challenges that are present while setting a remote holographic telepresence system? What are the basic requirements and their

Figure 3.
Holographic projection of an engineer professor from Zacatecas campus being projected in Monterrey campus.

Figure 4.
Holographic projection of professor Eduardo Luévano from Zacatecas campus being projected in Monterrey campus.

Figure 5.
Transparent acrylic mounted on an aluminum base.

classification that is needed for a live transmission? Is it possible to realize live transmissions in small clustered spaces? Is it possible to improve the communication by improving the visual perception of the image by correcting the parameters of the holographic telepresence?

6. Methodology

Action research refers to a wide variety of evaluative, investigative, and analytical research methods designed to diagnose problems or weaknesses—whether organizational, academic, or instructional—and help educators develop practical solutions to address them quickly and efficiently. Action research may also be applied to programs or educational techniques that are not necessarily experiencing any problems but that educators simply want to learn more about and improve [2].

Action research is a form of collective introspective inquiry undertaken by participants in social situations with the objective of improving the rationality and justice of their social practices or education, as well as the comprehension of these practices and the situations that they take place in.

It is a form of research that binds the experimental focus of social science with social action programs that respond to social principal problems. Because of social problems that emerge from the usual, action research starts the questioning of the phenomena from the usual, traveling systematically, as far as philosophical. Through action research what it is intended to treat in a simultaneous way are knowledge and social changes, in a manner that theory and practice unite.

The process of action research consists of:

1. Unsatisfaction with the current state of things (observe)

2. Identifying the problem area (observe)

3. Identifying a specific problem to be solved by action (think)

4. Formulation of multiple hypotheses (think)

5. Selecting a hypothesis (act)

6. Executing the action to prove the hypothesis (act)

7. Evaluation of the effects of the action

8. Generalizations.

Practitioners who engage in action research inevitably find it to be an empowering experience. Action research has this positive effect for many reasons. Obviously, the most important is that action research is always relevant to the participants. Relevance is guaranteed because the focus of each research project is determined by the researchers, who are also the primary consumers of the findings. Therefore, the proposed research methodology for this project will be of a qualitative nature, because it will allow the use of different sources of information and will produce descriptive data (peoples own words, written or spoken, and observable behavior). Its objective is the description of qualities of a phenomenon, since it seeks a deep understanding about the research topic. This methodology is of inductive character; it helps to understand the context and the people under a holistic perspective, that is, they are not reduced to variables, if not considered. It studies people in the context of their past and in the situations that they are in. Qualitative investigation is flexible as to how to conduct studies, it follows oriented guidelines, and its methods are at the service of the researcher, which is not dependent on a single procedure or technique. When using the qualitative methodology, one can obtain rich and profound information of which can draw inferences from data. It is conceived that qualitative methods are the first level of approach to reality, so that later a second level is possible to take on a more rigorous and profound methodology [13].

The qualitative methods to be used are interviews, action research, and observation. In some cases, surveys will be used as a quantitative method, with the sole purpose to enrich the investigation.

6.1 Observation

As a procedure of data recollection that allows to obtain information about a phenomena or event as it occurs. In some investigation processes where subjects are needed that cannot provide verbal information, observation is used as a data gathering method.

6.2 Interview

It is a technique in which a person solicits information from another one or from a group, to obtain data from a specific problem. It is believed to provide a deeper understanding of the experiment.

The quantitative methods to be used are:

6.3 Survey

It is a brief interview or discussion with individuals about a specific topic. It is a term often used to mean collect information. In this case, the survey is a list of questions aimed at extracting specific data from a specific group of people. The survey will be applied by email; a closed-ended questionnaire will be sent to the participants of the experiment.

7. I challenge

What happens when you put two minds in the same physical room? The exchange of ideas, talent, knowledge, and creativity. That is what holographic and telepresence technologies are going to enable. The level of engagement and interaction will become more human and meaningful.

In the search of making the telepresence experience the best possible, the Tecnológico de Monterrey research group has arrived at holographic projection. In this context of new technological resources, in the digital world of constant transformations, where this research group is currently working on, they developed an educational innovation that proposes holographic projections in real time, in which the professor can be seen and heard by the students, through a holographic projection that includes sound and voice in two ways.

Telepresence through holographic projection has been used in recent years as a manner of delivering conferences in international conferences. But toward giving official lectures in college level, there is only one record of initiatives on an experimental level. The telepresence with the holographic projection applied in a college course will allow the professor to, without limits as far as distance, weather, time difference, etc., give his class on time and in a form that while he is not physically there in the classroom. This technology enables cost saving because it will not be necessary to travel to other cities just to give a class, conference, or meeting. The use of the holographic projection is not only merely academic, it can be versatile and multipurpose in universities, for example, a directive that is out of town can attend a meeting through holographic projection [14].

In the University of Tecnológico de Monterrey, professor Eduardo Luévano from campus Zacatecas has been doing research and working on making the long-distance education process more efficient. His research is centered on the telepresence concept. He and his research team were searching to improve the telepresence sensation given by the professor, so they proposed to integrate a complement to long-distance education, which is holographic projection (**Figures 6** and 7). They believe that integrating this technology with videoconference and telepresence robot can assemble a technological package that will allow supplying, but never to replace, the temporary physical absence of the teacher in the classroom [15].

An initiative called "Reto i" (i challenge) is a collaboration network that was launched to a group of universities across all over Latin America that used multiple telepresence technologies ranging from traditional videoconference systems as Skype, telepresence robots, to a holographic display system. This was done to demonstrate that with the use of these technologies a professor or instructor can offer support or advice for multidisciplinary groups.

The designed instruments used were surveys, photographic recollection, and field notes. Some of the main results of the instruments were analyzed to prove the project impact:

- 87% perceived the holographic projection as the social presence of their professor.

- 86% of the students were satisfied with the project.

- 88% of the students felt comfortable with the "Professor Avatar."
- 93% would recommend this model to other students.
- 97% would participate again in telepresence projects.

Therefore, the engagement of the students is highly positive and productive.

Impact results:

The 1-week i challenge required students to construct a sustainable electric generator using recycled material found in dumpster and recycling centers that were useful to solve the necessity of a local community in poverty. Facing a real problem in a community promoted between the students a social commitment that allowed them to relate what they learned in a classroom to practice, work collaboratively, and to develop decision-making, communication, and leadership skills offer sponsorship for the project [15] .

Figure 6.
Black background for transmission and recording.

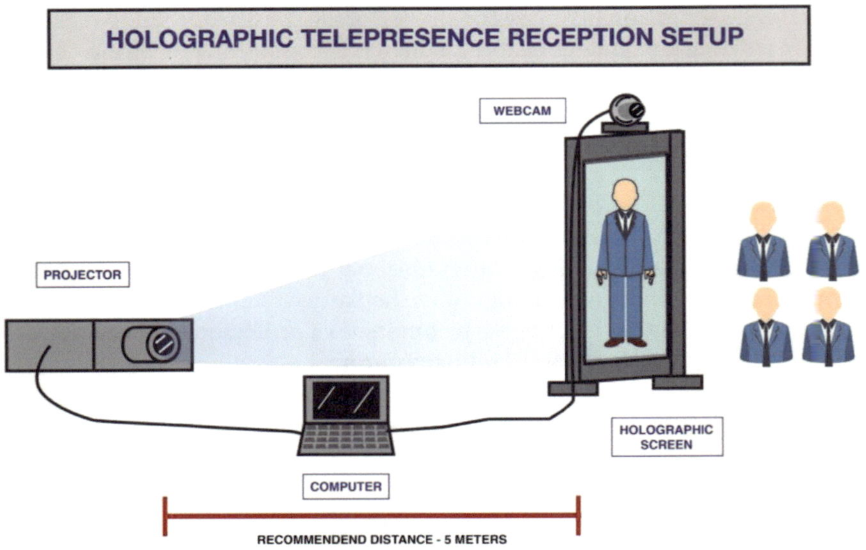

Figure 7.
Hologram reception setup.

Basic metrics to evaluate the students work in the 1-week i challenge were:

Energy generation (Watts-hour)	40%
Low cost (up to 100 USD)	40%
Social impact (people benefited)	20%

The student survey applied by the end of the project showed the following results:

100%	Considered that telepresence contributed to learning improvement.
87%	Considered that the activity goal was met.
97%	Considered that new skills were developed.
98%	considered that the i challenge helped to get them involved in their social, economic, and environmental reality.

The result of the project has five generators were constructed at full completion of the task requirements. By the end of i challenge, every institution donated the generator to a local vulnerable zone.

The scalability and potential of the project have motivated corporations as Samsung, BlackBoard, and Prezi to offer sponsorship for the project [16].

7.1 Setting up the remote holographic telepresence system

The key factors that define the correct setup for this means of communication are listed and described and divided into three categories: structure/components, software, and ambient considerations.

The essential structure and components for the parameters that are considered in this study are:

1. *Transparent glass or acrylic*. It is necessary to construct a projection screen in each site viewing the holograph. The screen is a pane of tempered glass or transparent acrylic, preferable 2 m high and at least 1.3 m wide. This transparent pane should be held upright in a sturdy manner and should have nothing immediately behind it.

2. *Holographic film*. This is a polarized semitransparent film sticker that adheres to a transparent glass or acrylic screen. The film has crystalline nanostructures that retain light emitted by the projector; this produces a holographic effect when the film is adhered to a glass or acrylic pane.

3. *Computer*. A computer is necessary for live transmission or for reproduction of recorded videos. The computer can be a desktop or laptop. A computer with more RAM memory and a faster processor will render greater image stability and quality.

4. *Internet connection*. For live transmission, it is critical to have a good internet connection, preferably 10 Mbps or greater. Connecting by a cable (Ethernet) is recommended over Wi-Fi because a cable generally provides a faster and more stable connection. If possible, we recommend designating an exclusive internet channel for transmission, and if this is not possible, it is recommended that participants in the session turn off the internet access of their mobile devices so that it will leave open bandwidth in the area.

5. *Projector*. Another key component is the projector. Here are some considerations:

Use a projector of at least 3500 lumens.

a. Long throw (standard) projectors emit uniform luminosity. When projecting an entire person in life-sized scale, the entire body is correctly illuminated. The restriction of long throw projectors is that they require a greater distance between the projector and the screen to achieve a life-size scale.

b. Short throw projectors have the advantage of reducing the distance necessary between the projector and screen to achieve life-size scale. The disadvantage of these machines is that they emit nonuniform luminosity. For example, when projecting an entire human at a life-size scale, only the upper half is well illuminated, while the legs fade out of sight. We recommend short throw projectors for half-body images. One example would be a person sitting at a desk and visible from the surface of the desk upward.

c. The projector is placed behind the holographic screen shining toward the audience, but at a slight angle so that its light does not hit the audience directly in the eyes. The holograph will appear uniformly visible to the audience.

6. *Webcams and cameras*. The quality of the holographic image depends on the webcam used by the presenter. For this reason, the recommendation of external HD webcams rather than those cameras integrated into laptops. Also an external webcam that is connected via the USB port can be more easily manipulated, and its position can be adjusted, so that the audience and the speaker can see either part with a better view.

7. *Audio equipment*. A small audio system (desktop speakers, Bluetooth speakers, surround sound, etc.) is necessary for transmitting voice and sound clearly to the entire audience. The dimensions of your audio system should correspond to the dimensions of your classroom or auditorium.

8. *Black background for transmission and recording*. The background can be a paper cyclorama or a fabric with no sheen, such as muslin. The black background is necessary during recording because it disappears during projection and leaves only the body of the professor in the holograph.

9. *Lighting equipment*. Professional lighting equipment is not necessary. It is necessary to fully illuminate the presenter, especially above the waist. Artificial light is preferable to natural light because it is controllable. Further, excessive light should be avoided because this causes shadows behind the professor or causes the backdrop to glow behind the professor. Lamps with dimmers can be useful for controlling the level of illumination. The ideal color temperature should be between 3000 and 3500 K [17].

10. When you light the background of the screen with artificial lighting with dimmers, it makes the hologram illusion even more believable.

The critical and essential software parameters that are considered in this study are:

1. *Videoconference software*. For this exercise, the videoconference that was used was Skype desktop app.

Ambient considerations:

1. *Strong daylight*. This issue is pertinent in any situations where projectors are being used as a source of video. Despite manufacturers having made great strides in the last couple of years with the brightness of their projectors, there will never be anything available that will be able to compete with natural daylight. In circumstances of strong natural daylight, the options are either to create a controlled light environment or make a stage with ceiling and walls.

2. *Space*. It is important to remember and to consider the transmission and the reception area. This is a key factor because it determines the conditions in which everything else is going to be set up.

For the transmission area, it is recommended that there should be sufficient space for at least an average adult to be standing in front of the black background so that the camera can capture the subject for a full body transmission. The recommended distance between the camera and the speaker should be between 1.5 and 2 m for a full body presentation.

For the reception area, it is recommended that there should be sufficient space for the audience. Here the factors depend on how many spectators are going to be, the size of the room or auditorium, etc. In the case that the projector being used is a long throw projector, it needs to be behind the holographic screen from a faraway distance, so that the image that is being projected can be adjusted to human scale. With short throw projectors, the distance from the holographic screen can be dramatically reduced, but using these projectors may reduce quality. For this purpose, it is deeply recommended set the reception of the transmission to test prior to a conference or lecture and to leave some marks of the position of everything.

3. *Dress code*. We recommend the presenter wear light colors because dark colors can be confused with the black background and disappear in the projection. Do not wear black. Avoid wearing elaborate patterns as these may project with a poor resolution depending on the quality of internet and other equipment.

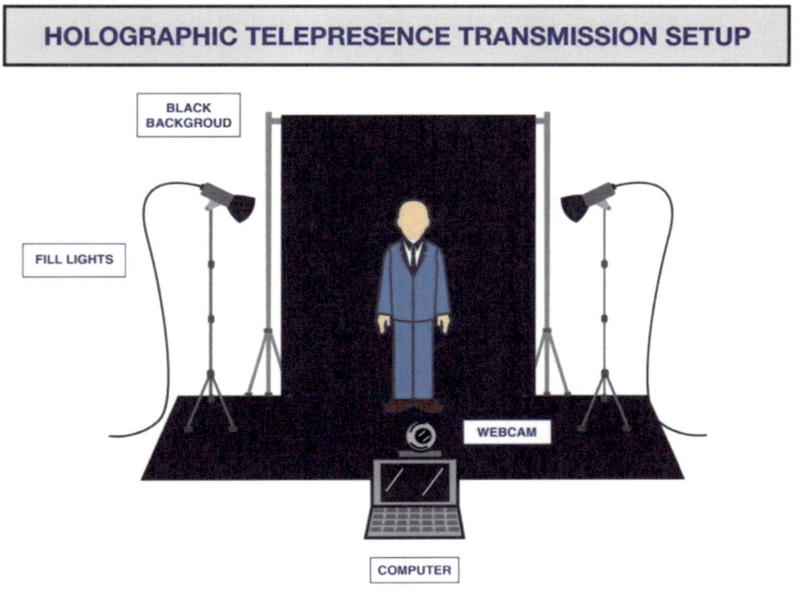

Figure 8.
Hologram transmission setup as seen in a front view.

In **Figure 7**, it is illustrated how the proposed remote holographic telepresence reception should be setup according to the requirements mentioned before. In this case, it will be used a long throw projector that is recommended to be set at 5 m behind the holographic screen, so that it will be a lot easier to adjust the image to a full human scale body (as illustrated in the figure). For the position of the webcam, the ideal position should be at the eye level of the spectators, but if that is not possible, another recommendation will be to put it on a location that gives it a broad field of view of the target audience, so that the speaker may be able to see who he is addressing.

In **Figure 8**, it is illustrated how the proposed remote holographic telepresence transmission should be setup according to the requirements mentioned before. In the figure, it is shown that the user needs to be in front of the black background. The position of the webcam needs to be adjusted so that the only thing visible that will be transmitted will be the black background and the user and nothing else. Depending on the space where the transmission will be taking place and the type of lighting, it may be necessary to have a couple of fill lights, so that the user may appear well lit.

8. Conclusions

It was verified that by identifying the parameters for a remote holographic telepresence transmission, challenges and complications could be identified; also it was possible to streamline and facilitate the assembly process, since this series of instructions served and could be replicated easily in another place without major problems. By correcting the parameters, it was possible to change drastically the quality of the image both recorded and live transmission.

As for the transmission, it was possible to do it in a reduced space of 0.5–2 m of distance of the camera, although only the projection left in half body. But for the reception of the holographic projection due that the work was made with a long throw projector the most that the space could be reduced space sacrificing human scale was 4 m of distance between the projector and the holographic screen. Tests with short throw projectors were not made due to the fact that the equipment constantly failed and overheated which was useless to test on.

3D holographic projection technology clearly has a big future ahead. As this audiovisual display continues to get high profile credibility, we are likely to see more companies advertising their products or marketing their business in this way. Holographic telepresence can revolutionize the way we think about and experience modern communication systems. In fact, it has the potential to change diverse types of communication systems.

Holographic projectors will be able to render sharp projected images from relatively small projection devices (e.g. cell phones) because they do not require high intensity, high-temperature light sources. Researchers at different industries and schools are working toward applied science that could make real-time holographic projections in everyday-used devices [8].

Acknowledgements

I wish to thank Dr. Eduardo González Mendívil for opening the doors to the research group and providing constant support in this journey and the members of the Thesis Committee, M.Sc. Pablo Guillermo Ramírez Flores, and Dra. Norma Patricia Salinas Martínez, for having shared projects together helping to develop my skills in each of these.

To my team members and colleagues M.Sc. Gabriel Pantoja García, Engineer Héctor Eduardo Ramírez, professor Luis Eduardo Luévano Belmonte, with whom I collaborated in different virtual and augmented reality and holographic telepresence projects. As well all the people involved in Virtual and Augmented Reality in Education (VARE 2015) conference and in the RWTH-Tecnológico de Monterrey holographic telepresence transmission.

Author details

Luis Luevano*, Eduardo Lopez de Lara and Hector Quintero
Tecnologico de Monterrey, Monterrey, Zacatecas, Mexico

*Address all correspondence to: luevano@tec.mx

References

[1] Minsky M. Marvin minsky's telepresence manifesto. 2010. Retrieved from IEEE Spectrum: http://spectrum.ieee.org/robotics/artificial-intelligence/telepresence-a-manifesto

[2] McKernan J. Curriculum Action Research. A Handbook of Methods and Resources for the Reflective Practitioner. London: Kogan; 1996

[3] Steinmeyer J. Hiding the Elephant. New York: Carroll & Graf; 2003

[4] Capucci PL. The case of holography among media studies, art and science. Technoetic Arts: A Journal Of Speculative Research. 2011;**9**(2/3):247-253

[5] SUMMIT. Introducing holographic telepresence: Is it the future? 2015. Retrieved from: http://www.summit-sys.com/, http://www.summit-sys.com/blog/introducing-holographic-telepresence-future/

[6] Bove VM. Engineering for live holographic TV. SMPTE Motion Imaging Journal. 2011;**120**(8):56-60

[7] Rodiguez T, Cabo de Leon A, Uzzan B, Livet N, Boyer E, Geffray F, et al. Holographic and Action Capture Techniques. SIGGRAPH '07 ACM SIGGRAPH 2007 Emerging Technologies. New York: ACM; 2007

[8] Elmorshidy A. Holographic projection technology: The world is changing. Journal of Telecommunications. 2010;**2**(2):104-112

[9] Roger Y. Holograms could Give Virtual Meetings New Life. USA Today; 2011

[10] Cohen A. Holographic videoconferencig. Futurist. 2011;**45**(3):14-15

[11] Stile L. 3D breakthrough with updatable holographic displays. Space Daily. 2008

[12] Arjona K. Hologramas, un paso más en la presencia virtual. 2014. Retrieved from: www.calidadytecnologia.com, http://www.calidadytecnologia.com/2014/05/hologramas-tele-presencia-Musion-Cisco.html

[13] McNiff J. Action research for professional development. 2002. Retrieved from: http://www.jeanmcniff.com/, http://www.jeanmcniff.com/ar-booklet.asp

[14] Luévano E, López de Lara E. Uso de dispositivo móvil de telepresencia en la educación a nivel universitario. In: Congreso Iberoamericano de Ciencia, Tecnología, Innovación y Educación. Buenos Aires, Argentina: OEI; 2014. pp. 1-12

[15] Luévano E, López de Lara E, Castro J. Use of telepresence and holographic projection mobile device for college degree level. Procedia Computer Science. 2015;**75**:339-347

[16] Luévano E. 2016. www.profesoravatar.com. Retrieved from Professor Avatar: https://profesoravatar.com/

[17] Lichtman HS. Creating telepresence environments. 2011. Retrieved from Telepresence Options: http://www.scribd.com/doc/58471628/Creating-Telepresence-Environments

[18] Barras C. 'Holographic' videoconferencing moves nearer to market. 2009. Retrieved from New Scientist: https://www.newscientist.com/article/dn18169-holographic-videoconferencing-moves-nearer-to-market/

Section 3

Algorithms and Image Processing

Chapter 7

Real-Time Diffraction Field Calculation Methods for Computer-Generated Holograms

Gokhan Bora Esmer

Abstract

Holographic three-dimensional television systems provide a natural 3D visualization. Fast calculation of the diffraction field from a three-dimensional object is essential to achieve video rate. In the literature, there are myriads of fast algorithms for diffraction field calculation from three-dimensional objects, but most of them omit the pixelated structure of the dynamic display devices which are used in the reconstruction process. In this chapter, the look-up table-based fast algorithm for diffraction field calculation from a three-dimensional object for a pixelated dynamic display device is presented. Real-time diffraction field calculations are obtained by running the algorithm in parallel on a graphical processing unit. Performance of the algorithm is evaluated in terms of computation time of the diffraction field and the normalized mean square error on the reconstructed object. To have optimization on the required memory space for the look-up table, two different sampling policies along the longitudinal axis are implemented. Uniform sampling policy along the longitudinal axis provides better error performance than nonuniform sampling policy. Furthermore, optical experiments are performed, and it is observed that both numerical and optical reconstructions are similar to each other. Hence, the proposed method provides successful results.

Keywords: computer-generated holograms, holographic display, real-time holography, spatial light modulators, 3D visualization

1. Introduction

Holography is the only visualization technique that satisfies all the depth cues [1–3]. Therefore, it gives a natural three-dimensional (3D) visualization. Holography is based on capturing the diffracted optical waves from an object and regenerating those waves again by illuminating the recording media [1–6]. Captured optical waves provide a significant amount of information related to the object such as surface profile, depth, and refractive index of the object. Hence, holography has a myriad of applications. For instance, holograms can be used as optical elements like prisms, lenses, and mirrors [7, 8]. Also, parallel optical computing is possible when holograms are employed [9, 10]. Furthermore, holograms are useful in metrology [11–13] and microscopic imaging to visualize very small

IntechOpen

objects like cells and bacterias [14, 15]. Another application of holography is related to nondestructive testing [16–18]. Nevertheless, major application of holography is related to 3D visualization, and it is used in education [19, 20], dentistry [21, 22], gaming [23], demonstration of cultural heritage [24], and more.

Holography setups can be assembled by using different configurations depending on the application. In optical holography setups, holographic patterns are stored on high-resolution holographic films [25, 26] and some type of crystals [27]. However, in some of the applications, we need to process the captured holographic patterns by numerical methods. Then, digital sensing devices are employed as a capturing device. Those types of setups are called as digital holography, and it has a vast amount of applications especially in nondestructive testing and microscopy. In [28], digital holography-based measurement method of 3D displacement is presented. Observed material is illuminated from four different directions sequentially; then they are combined to improve the resolution in the order of 10 nm. As a nondestructive testing method, digital holography is used in the analysis of cortical bone quality and strength impact in [29]. Furthermore, a method based on digital holography is implemented for detecting and measuring effect of moisture on the hygroscopic shrinkage strain on wood [30]. Another application of digital holography is in precise and accurate measurement of the initial displacement of the canine and molar in human maxilla [31]. By using subpixel registration and fusion algorithms, an improvement of profile measurements and expanding the field of view (FOV) in continuous-wave terahertz reflective digital holography is achieved [32]. A comprehensive review of denoising methods on phase retrieval from digital holograms in terms of signal-to-noise ratio (SNR) and computation time is presented in [33]. Removal of phase distortions by using principal component analysis (PCA) method is given in [34].

Holography is a versatile tool for visualization, measurement, and testing. In optical and digital holography methods, we need some optical sensing elements like polymers and digital devices to capture the diffracted field from the object. However, in computer-generated holography (CGH), diffraction field calculations are performed by using numerical methods and signal processing algorithms [4–6, 35]. Then, we can obtain the hologram from the calculated diffraction field and use it to drive dynamic display devices such as spatial light modulators (SLMs). After that, illumination of the SLM with a coherent light source will provide an optical reconstruction of the original object. When CGHs are calculated sequentially and used in driving SLMs, then we can have a holographic 3D television (H3DTV) as a product. An overview on holographic displays is presented in [36]. Generally, coherent light sources are used in H3DTV systems, and those light sources can generate speckle noise in the reconstructions. Low computational method for improving image quality and decreasing the speckle noise in CGH is proposed in [37]. Diffraction field calculations as in CGH are also used in other 3D display systems to improve the resolution of reconstructed objects. For instance, in integral imaging-based 3D display system, distortions on the elemental images are corrected by using holographic functional screen [38].

In diffraction field calculation from a 3D object, we have to generate a synthetic 3D object. There are plenty of ways for generating a synthetic 3D object in a computer. For instance, we can form a 3D object by using a set of point light sources which are distributed over the space. Those types of objects are called as point cloud objects. To calculate the diffraction field from a point cloud object, we superpose the diffraction fields emitted by each point light source [35, 39–44]. Another 3D object generation method is based on stitching small planar patches. As in the process of diffraction field calculation from point cloud objects, once again the diffracted fields from each patch are superposed to obtain the diffraction field of the object [45–55]. The third method which can be used in the generation of synthetic 3D object is based on having multiple two-dimensional (2D) cross sections of the object along the longitudinal axis. Then, superposition of diffracted fields from those 2D cross sections will give the diffraction field of the 3D object [56–60]. A detailed summary on CGHs in terms of resolution, field of view, eye relief, and optical setups for different 3D object generation methods can be seen in [61, 62].

CGHs of the objects should be calculated rapidly to obtain H3DTV systems. Hence, fast methods such as fast Fourier transform (FFT) and look-up table (LUT)-based methods are utilized in CGH calculations. In [39, 52], algorithms which are based on FFT are used for decreasing the calculation time of CGH. Precomputed LUTs are also used for achieving fast calculations in CGH calculations [2, 39, 41, 42, 63, 64]. Another way to achieve fast calculation in CGH is based on segmentation of diffraction field from point light sources [43, 44]. Parallel processing of diffraction field calculation provides further improvements on the computation time. Graphical processing units (GPUs) are special hardware to run parallel calculations. Thus, they are one of the most convenient hardware for H3DTV systems [40, 44, 65, 66]. Time-division method can also be used in the calculation of CGHs for layered 3D objects to achieve fast computations [67].

Imposing some approximations in the diffraction field calculations provides to decrease the computational complexity, and it paves the way to obtain fast diffraction field calculations. In the meantime, we have to improve the quality of the reconstructed object. An accurate calculation method of diffraction field which is based on angular spectrum decomposition is explained in [68]. Furthermore, diffraction field calculation methods for SLMs with pixelated structure are presented in [69–71]. However, the computational complexities of those methods are too high to have real-time diffraction field calculations. As a result of this, the algorithms presented in [72–75] are proposed as a solution to both computation time and quality in the reconstructed object in H3DTV. Further computational time improvements can be obtained by utilizing a LUT which is optimized for parallel processing on a GPU to achieve real-time calculations. Moreover, the pixel structure of the employed SLM in the reconstruction process is taken into account in forming LUT. Calculated LUT has one-dimensional (1D) kernels to decrease the allocated memory space.

2. Calculation of diffraction pattern used in driving SLM with pixelated structure

In CGH, it is possible to obtain 3D reconstructions of both synthetic and real objects. By employing dynamic display devices like SLMs in the reconstruction process, we can have H3DTV systems. To drive SLMs, we have to calculate diffraction fields from 3D objects by using numerical analysis methods and signal processing techniques. Calculation of diffraction field depends on the 3D object generation method. In this work, we assumed that 3D objects are represented as point clouds, because it is one of the simplest methods in 3D object generation. The diffraction field of the 3D object is calculated by superposing the diffraction fields emitted from the points that form the 3D object.

Superposition on diffraction field calculation from a point cloud object over a planar surface can be expressed as

$$\psi(\boldsymbol{r}_0) = \sum_{l=1}^{L} \psi(\boldsymbol{r}_l) h_F(\boldsymbol{r}_0 - \boldsymbol{r}_l) \tag{1}$$

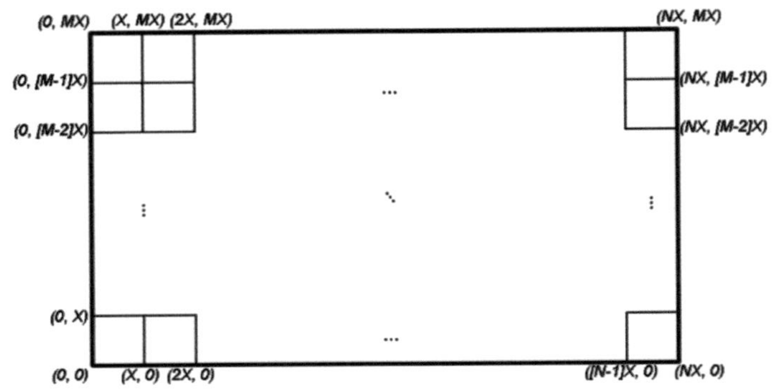

Figure 1.
An illustration of the pixel structure of the simulated SLM.

where $\psi(\boldsymbol{r}_0)$ and $\psi(\boldsymbol{r}_l)$ are the diffraction fields over SLM and diffraction field at l^{th} sample point of the 3D object, respectively. Surface of SLM is represented by the position vector $\boldsymbol{r}_0 = [x, y, 0]$, and the sampling points of 3D object are shown by $\boldsymbol{r}_l = [x_l, y_l, z_l]$. We assume that Fresnel approximation is valid and the term $h_F(\boldsymbol{r})$ denotes diffracted field on the SLM from a point light source expressed as

$$h_F(\boldsymbol{r}) = \frac{e^{jkz}}{j\lambda z} e^{\frac{jk}{z}(x^2+y^2)} \tag{2}$$

where $\boldsymbol{r} = [x, y, z]$, k is the wave number, and λ is the wavelength of the light source used in illumination of the object.

Scaled and superposed diffraction fields from point light sources provide the diffraction field of the 3D object, and its phase component is used for driving the SLM. Then, entire surface of the SLM is illuminated by a plane wave. After that, the reflected optical wave from the surface of the SLM generates an optical replica of the 3D object. Most of the off-the-shelf SLMs have square pixels with very high filling factors like 93% [76]. Hence, the filling factor in the simulated SLM is approximated as 100%. The pixel structure of the simulated SLM is illustrated in **Figure 1**.

Simulation of optical setup can be improved when the pixelated structure of the SLM is taken into consideration. For that purpose, we have to perform surface integration over each pixel area on the SLM. It is assumed that gray value over each pixel area has a constant value. The diffraction field over SLM can be found as

$$\psi_{2D,z=0}(n,m) = \int_{x_n}^{x_{n+1}} \int_{y_m}^{y_{m+1}} \sum_{l=1}^{L} \psi(\boldsymbol{r}_l) h_F(\boldsymbol{r}_0 - \boldsymbol{r}_l) dx dy \tag{3}$$

where n and m stand for indices of SLM along x- and y-axes, respectively. It is also possible to represent Eq. (3) by scaling and superposing 2D kernels $\boldsymbol{K}_{\alpha_l, 2D}$,

$$\psi_{2D,z=0} = \sum_{l=1}^{L} P(\boldsymbol{r}_l) \boldsymbol{K}_{\alpha_l, 2D} \tag{4}$$

where $P(\boldsymbol{r}_l) = -j\psi(\boldsymbol{r}_l)e^{jkz_l}$ and $\alpha_l = \frac{1}{\sqrt{\lambda z_l}}$ 2D kernel $\boldsymbol{K}_{\alpha_l, 2D}$ can be decomposed into 1D kernels as

$$\boldsymbol{K}_{\alpha_l,\,2D} = \left(\boldsymbol{K}^{x_l}_{\alpha_l,\,1D}\right)^T \boldsymbol{K}^{y_l}_{\alpha_l,\,1D} \tag{5}$$

where x_l and y_l refer to locations of lth point light source, used in generation of 3D object, along x- and y-axes, respectively. Each 1D kernel $\boldsymbol{K}_{\alpha_l,\,1D}$ can be represented as

$$\boldsymbol{K}_{\alpha_l,\,1D} = \left[K_{\alpha_l,\,1D}(1)\right\} \;K_{\alpha_l,\,1D}(2) \;\;\cdots\;\; K_{\alpha_l,\,1D}(N)\,] \tag{6}$$

and its elements can be calculated as

$$K_{\alpha_l,\,1D}(n) = C\left(\zeta_{l,\,n+1}\right) + jS\left(\zeta_{l,\,n+1}\right) - C\left(\zeta_{l,\,n}\right) - jS\left(\zeta_{l,\,n}\right) \tag{7}$$

where $\zeta_{l,\,n} = \frac{x_n - x_l}{\sqrt{\lambda_{z_l}}}$. The operators $C(\cdot)$ and $S(\cdot)$ stand for cosine and sine Fresnel integrals, respectively [5, 6], and they are calculated as

$$C(w) = \int_0^w \cos\left(\frac{\pi}{2}\tau^2\right) d\tau;\ \mathrm{S(w)} = \int_0^w \sin\left(\frac{\pi}{2}\tau^2\right) d\tau \tag{8}$$

Numerical evaluation of cosine and sine Fresnel integrals given in Eq. (8) is calculated by adaptive Lobatto quadrature [77].

In the standard algorithm, diffraction field of each point is obtained by evaluating Eq. (8). Then, superposition of those fields is performed to obtain CGH. As a result of this, computational complexity of diffraction field calculation is too high to have real-time applications. As a solution to the computation time problem, we present a fast algorithm to calculate 2D kernel, $\boldsymbol{K}_{\alpha_l,\,2D}$, based on LUT and parallel processing.

3. Proposed algorithm for fast calculation of CGH

Fast computation of diffraction field and improved quality of reconstructed 3D object are essential issues in H3DTV. As a solution to those problems, we propose a method based on calculation of 2D kernels $\boldsymbol{K}_{\alpha_l,\,2D}$ without evaluating sine and cosine Fresnel integrals. To achieve fast calculation, precomputed LUT is utilized, and the diffraction field of the 3D object can be obtained by scaling and superposing the 2D kernels $\boldsymbol{K}_{\alpha_l,\,2D}$ as

$$\hat{\boldsymbol{\psi}}_{2D,\,z=0} = \sum_{l=1}^{L} \boldsymbol{P}(\boldsymbol{r}_l)\hat{\boldsymbol{K}}_{\alpha_l,\,2D} \tag{9}$$

where $\hat{\boldsymbol{\psi}}_{2D,\,z=0}$ denotes the estimated diffraction field of the 3D object on the SLM and $\hat{\boldsymbol{K}}_{\alpha_l,\,2D}$ is the 2D kernel which denotes the diffraction field of $\boldsymbol{l}^{th}$ point of the 3D object on SLM. 2D kernel $\hat{\boldsymbol{K}}_{\alpha_l,\,2D}$ is calculated by multiplying 1D kernels $\boldsymbol{K}_{\alpha_l,\,1D}$ from LUT as shown in Eq. (5). Each 1D kernel $\boldsymbol{K}_{\alpha_l,\,1D}$ represents the diffraction field on SLM from specific depth along longitudinal axis. A simple arithmetic operation is used for speeding up data fetching from the LUT. As result of this, total computation time of the diffraction field can be improved in terms of data fetching. By increasing the number of precomputed 1D kernels in LUT, we can achieve better diffraction field estimations for proposed method, but it causes to allocate more memory space. Hence, we apply different sampling policies along longitudinal axis to optimize memory space allocation. In the first sampling policy, uniform sampling along longitudinal axis is performed. In the second sampling policy, we sample the parameter $\alpha_l = \frac{1}{\sqrt{\lambda z_l}}$ uniformly. Thus, we have nonuniform sampling along the longitudinal axis.

4. Simulation results

Performance assessment of the proposed diffraction field calculation method is obtained by implementing different scenarios, but a few of them are presented to give an insight to the reader. Two major performance evaluation criteria are taken into account: total computation time of the CGH and the normalized mean square error (NMSE) on the reconstructed object. NMSE on the reconstructed object can be calculated as

$$NMSE = \frac{\sum_{n=1}^{N}\sum_{m=1}^{M}\left|\hat{\psi}_{2D,z=z_0}(n,m) - \psi_{2D,z=z_0}(n,m)\right|}{\sum_{n=1}^{N}\sum_{m=1}^{M}\left|\psi_{2D,z=z_0}(n,m)\right|} \tag{10}$$

where $\psi_{2D,z=z_0}(n,m)$ and $\hat{\psi}_{2D,z=z_0}(n,m)$ denote reconstructed objects at $z = z_0$ plane from the diffraction field calculated by the standard and the proposed algorithms, respectively. Simulated scenario for a CGH is illustrated in **Figure 2**.

First, a 3D point cloud object is generated in computer environment. The generated 3D object has **3144** points which are distributed over the space. The volume occupied by the object has $x_e = 2.8mm$, $y_e = 4.1mm$, and $z_e = 4.1mm$ extensions along x-, y-, and z-axes. There is a distance between the object and the screen, and it is taken as $z_0 = 61.6mm$. We assume that simulated SLM has **100%** fill factor and pitch distance X_s is taken as **8 μm**. Also, the simulated SLM has **512** pixels along both x- and y-axes, respectively. We assume that green laser is employed for illumination purpose; hence the wavelength is taken as **532 nm**.

The proposed algorithm is implemented by using two platforms: MATLAB and Visual C++. To have shorter computation time for diffraction fields, we utilize CUDA libraries and parallel computation power of GPU. The assembled computer system has i5-2500 CPU at 3.3 GHz, 4GB RAM, and a GTX-680 GPU to run the algorithm. Operating system of the computer is chosen as 64-bit Windows 7.

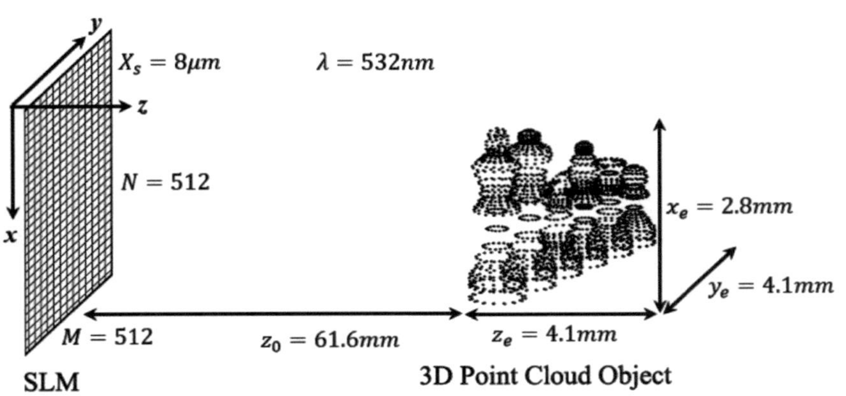

Figure 2.
An illustration of simulated optical setup. The SLM employed in the setup has **N** *and* **M** *pixels along* x*- and* y*-axes, respectively. Transversal axis sampling is indicated by* X_s*. The variable* z_0 *determines the distance between SLM and the closest point light source of the 3D object.*

Generally, off-the-shelf SLMs have pixelated structure, and phase parts of the calculated diffraction fields are used for driving the SLM. When the pixelated structure of SLM is not taken into account in CGH calculations, it is not easy to differentiate focused and unfocused parts of the reconstructed 3D objects. An illustration of such a result can be seen in **Figure 3a**. As a result of the similarity in focused and unfocused parts, the quality of the reconstructed object is decreased significantly. On the contrary, the difference between focused and unfocused parts in the reconstructed 3D object is clear when the proposed method is used in diffraction field calculation. Those results can be seen easily in **Figure 3b**.

Furthermore, numerical and optical reconstructions are very similar to each other, and that similarity in the reconstructions can be seen in **Figure 4**.

To calculate the diffraction field in the standard method, we need to perform cosine and sine Fresnel integrals for each pixel on SLM and for each point light source in 3D object. As a result of this, computational complexity of the standard method is extremely high, and CGH is calculated at **2701.10 s**. Significant improvement in computation time can be achieved when the proposed algorithm is

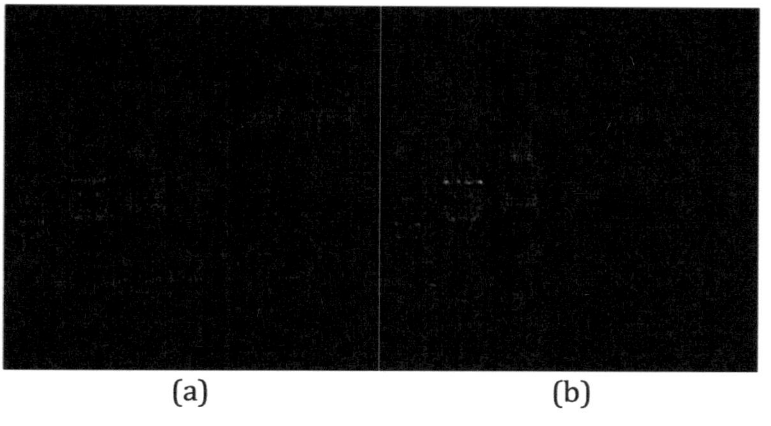

Figure 3.
A point cloud object which has six parts and each part is located at different depths along the longitudinal axis. The leftmost piece is reconstructed in both of the figures shown above: (a) reconstruction of the 3D object from the CGH obtained without taking into consideration the pixelated structure of SLM and (b) reconstruction from the CGH calculated by the proposed algorithm.

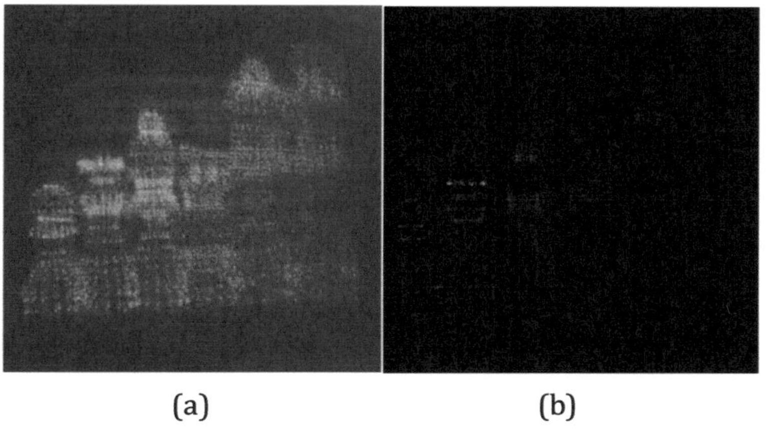

Figure 4.
(a) Optical reconstruction of a point cloud object and (b) numerical reconstruction of the same object given in (a).

employed in CGH calculation. When we use LUT-based method for the same scenario which is mentioned above, we need **8.15 s** to calculate the CGH. Further improvement in computation time can be obtained if the presented algorithm is implemented in parallel on a GPU. Although, significant gain on the computation time of CGH is obtained by using LUT, there will be negligible amount of error on the reconstructed objects, because of having finite number of kernels and the quantization effect along the longitudinal axis. The performance of the proposed method is summarized in **Table 1**.

By increasing the number of kernels in LUT, we can improve error performance of the algorithm without having any extra computational load, but there is an increase in the size of the required memory. As a result of this, installed memory space may not be enough to perform the diffraction field calculations. To overcome memory allocation problem, we use another sampling policy in generation of LUT. Two different sampling policies along the longitudinal axis are proposed. The first sampling policy is based on uniform sampling of longitudinal axis. The second sampling policy is related to uniform sampling of α_l. Hence, there will be nonuniform sampling along the longitudinal axis. **Tables 2** and **3** summarize performances of the sampling policies in terms of NMSE and required memory allocation by the precomputed LUT. As it can be seen from **Tables 2** and **3**, when the size

3D object = 3144 points; $\lambda = 532$ nm; N = M = 512; $X_s = 8$ µm	Computation time (s)	NMSE
Standard method	2710.10	—
LUT	8.15	0.08
LUT: parallel processing by using four cores	7.08	0.08
LUT: parallel processing by using GTX-680	0.08	0.08

Proposed algorithm utilizes LUT which has 1D precomputed kernels for 125 different sampling points along longitudinal axis.

Table 1.
Performance assessment of the proposed algorithm in terms of NMSE.

3D object = 3144 points; $\lambda = 532$ nm; N = M = 512; $X_s = 8$ µm		
Number of 1D kernels	**NMSE**	**Memory allocation (kB)**
83	0.068	332
92	0.061	368
103	0.054	412
118	0.048	472
137	0.039	548
165	0.034	660
206	0.026	824
274	0.020	1096
411	0.014	1644
821	0.006	3284

LUT is formed by uniform sampling of depth parameter along longitudinal axis. Each element in 1D kernels is represented by four bytes.

Table 2.
Performance of the proposed algorithm according to the number of kernels used in LUT, NMSE, and allocated memory space.

of the LUT is fixed, uniform sampling policy along longitudinal axis provides better NMSE performance than nonuniform one.

In terms of calculated numerical errors, there should be a significant amount of deviation between reconstructed objects from CGHs obtained by standard and proposed method, but it is not easy to differentiate the reconstructions visually. Illustrations of numerically reconstructed objects by using both methods are shown in **Figure 5a** and **b**, respectively. To see the difference between to reconstructions, we subtract two reconstructions from each other and then take the magnitude of that difference. Then, we scale difference image linearly between 0 and 255 to improve the visibility of insignificant deviations. Those deviations can be seen in **Figure 5c**. Most of the deviations are in the unfocused region, and those deviations will not decrease the quality of the reconstruction. As a result of this, the proposed algorithm provides successful results.

3D object = 3144 points; $\lambda = 532$ nm; N = M = 512; $X_s = 8$ μm		
Number of 1D kernels	**NMSE**	**Memory allocation (kB)**
83	0.127	332
92	0.114	368
103	0.104	412
118	0.088	472
137	0.077	548
165	0.062	660
206	0.051	824
274	0.038	1096
411	0.025	1644
821	0.013	3284

LUT is formed by uniform sampling of α_l parameter. Each element in 1D kernels is represented by four bytes.

Table 3.
Performance of the proposed algorithm according to the number of kernels used in LUT, NMSE, and allocated memory space.

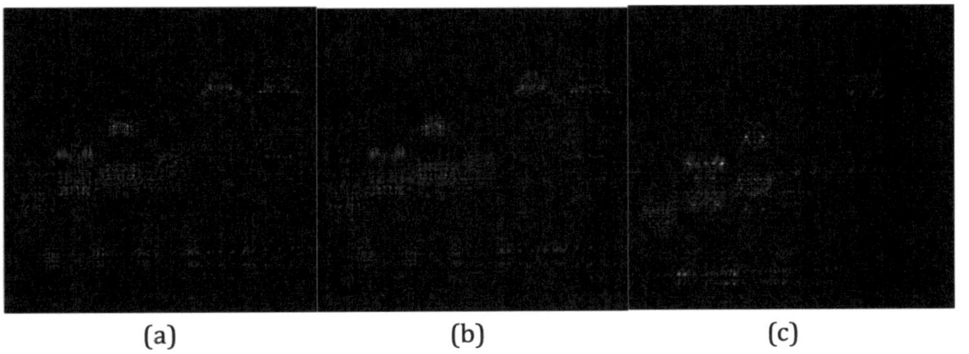

Figure 5.
(a) Magnitude of the reconstructed object at $z = z_0$ from the diffraction pattern calculated by standard algorithm and (b) by the proposed algorithm. (c) Magnitude of the difference between the reconstructed objects given in (a) and (b). Please note that image is scaled linearly from 0 to 255; thus the insignificant differences may become visible.

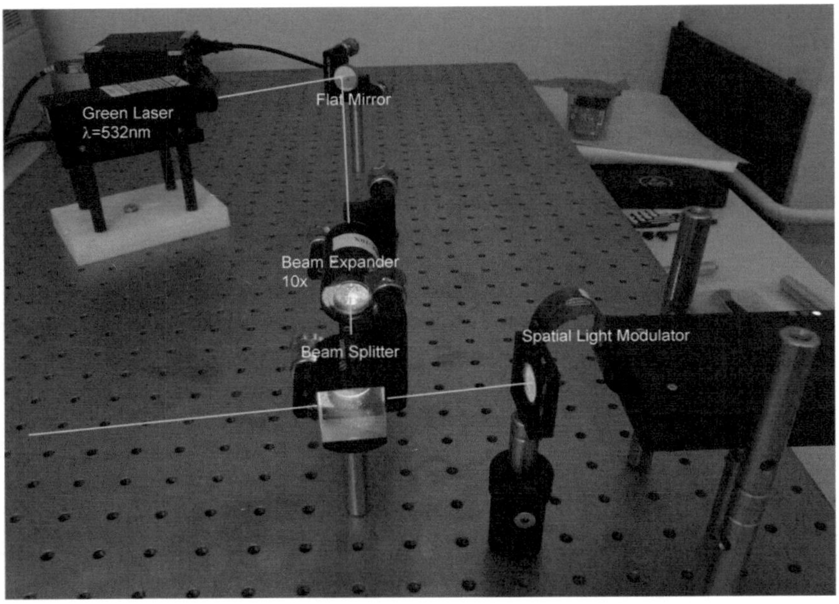

Figure 6.
Assembled optical setup for optical experiments.

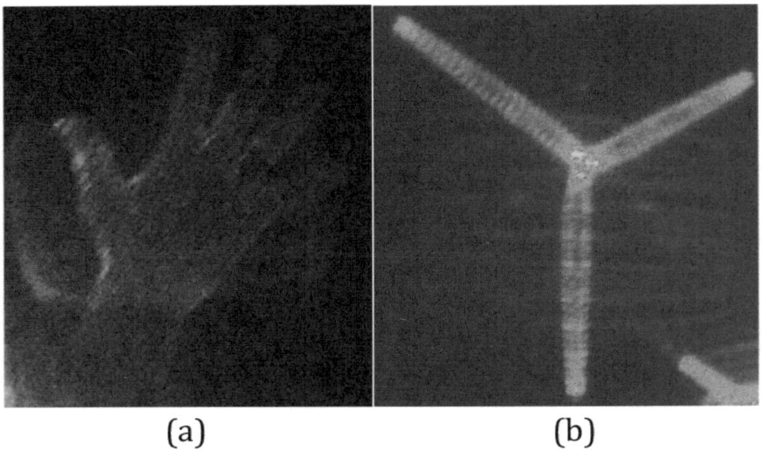

Figure 7.
Optically reconstructed 3D objects: (a) hand (b) propeller.

Performance assessment of the presented algorithm is tested by optical reconstructions as well. For that purpose, we assembled an optical setup which is shown in **Figure 6**. Green laser with $\lambda = 532$ **nm** is used as a coherent light source, and HoloEye Pluto phase-only SLM is employed as a dynamic display device. A couple of optically reconstructed objects are shown in **Figure 7**.

5. Conclusions

Two major problems in H3DTV systems can be called as decreasing the computation time of CGH and improving the quality of the reconstructed object. Using fast algorithms in diffraction field calculations will be helpful to decrease the computation time, but most of those fast algorithms impose some approximations that decrease the quality of the reconstructed object. In this work, we propose a diffraction field calculation algorithm that paves the way to achieve real-time calculations of diffraction fields from point cloud objects. In the meantime, the quality of the reconstructed objects is improved by taking into account the pixelated structure of SLM. Also, the proposed method can be run in parallel on a GPU. Performed numerical and optical experiments provide similar results. The proposed method utilizes precomputed LUT to decrease the computational load. To store the precomputed LUT, we need significant amount of memory allocation, and optimization of the occupied memory space is obtained by having two different sampling policies along the longitudinal axis. In the first sampling policy, LUT is formed by having uniform sampling along longitudinal axis. In the second one, nonuniform sampling is applied. When we fix size of the LUT, better NMSE performance is obtained by uniform sampling policy. As a result of this, when we use uniform sampling policy in computation of LUT, we need to allocate less amount of memory to store it.

Acknowledgements

This work was supported by the Scientific and Technological Research Council of Turkey project under grant EEEAG-112E220 and Marmara University Scientific Research Fund project under grant FEN-A-130515-0176.

Author details

Gokhan Bora Esmer
Marmara University, Istanbul, Turkey

*Address all correspondence to: bora.esmer@marmara.edu.tr

References

[1] Saxby G. Practical Holography. 3rd ed. Boca Raton, FL: Taylor and Francis; 2003. 478 p. ISBN: 978-1-4200-3366-3

[2] Lucente M. Diffraction-specific fringe computation for electro-holography [thesis]. Cambridge, MA: Massachusetts Institute of Technology; 1994

[3] Benton SA, Bove VM Jr. Holographic Imaging. New Jersey: John Wiley & Sons; 2008. 288 p. ISBN: 978-0470068069

[4] Toal V. Introduction to Holography. US: CRC Press Taylor and Francis Group; 2012. 502 p. ISBN: 978-1439818688

[5] Goodman JW. Introduction to Fourier Optics. 2nd ed. New York: McGraw Hill; 1996. 441 p. ISBN: 978-0070242548

[6] Born M, Wolf E. Principles of Optics: Electromagnetic Theory of Propagation, Interference and Diffraction of Light. 7th ed. New York: Cambridge University Press; 1980. 952 p. ISBN: 978-0521642224

[7] Hong K, Yeom J, Jang C, Hong J, Lee B. Full-color lens-array holographic optical element for three-dimensional optical see-through augmented reality. Optics Letters. 2014;**39**(1):127-130. DOI: 10.1364/OL.39.000127

[8] Wang Q-H, He M-Y, Zhang H-L, Deng H. Augmented reality 3D display system based on holographic optical element. In: Proceedings Volume 10942, Advances in Display Technologies IX; 1094203. SPIE OPTO; 2019. DOI: 10.1117/12.2508136

[9] Georgiou A, Kollin JS, Diaz AG. Multi-beam optical system for fast writing of data on glass. In: United States Patent US10181336B1. 2019

[10] Horst F. Integrated optical circuit for holographic information processing. In: United States Patent US20190041796A1. 2019

[11] Wagner C, Seebacher S, Osten W, Jüptner W. Digital recording and numerical reconstruction of lensless Fourier holograms in optical metrology. Applied Optics. 1999;**38**(22):4812-4820. DOI: 10.1364/AO.38.004812

[12] Picart P, Mounier D, Desse JM. High-resolution digital two-color holographic metrology. Optics Letters. 2008;**33**(3):276-278. DOI: 10.1364/OL.33.000276

[13] Claus D, Pedrini G, Buchta D, Osten W. Accuracy enhanced and synthetic wavelength adjustable optical metrology via spectrally resolved digital holography. Journal of the Optical Society of America A. 2018;**35**(4):546-552. DOI: 10.1364/JOSAA.35.000546

[14] Kim MK. Digital Holographic Microscopy: Principles, Techniques and Applications. New York: Springer Series in Optical Sciences; 2011. ISBN: 978-1-4419-7792-2

[15] Quan X, Kumar M, Matoba O, Awatsuji Y, Hayasaki Y, Hasegawa S, et al. Three-dimensional stimulation and imaging-based functional optical microscopy of biological cells. Optics Letters. 2018;**43**(21):5447-5450. DOI: 10.1364/OL.43.005447

[16] Gholizadeh S. A review of non-destructive testing methods of composite materials. Procedia Structural Integrity. 2016;**1**:50-57. DOI: 10.1016/j.prostr.2016.02.008

[17] Kreis T. Application of digital holography for nondestructive testing and metrology: A review. IEEE Transactions on Industrial Informatics.

2016;**12**(1):240-247. DOI: 10.1109/TII.2015.2482900

[18] Karray M, Christophe P, Gargouri M, Picart P. Digital holographic nondestructive testing of laminate composite. Optical Engineering. 2016; **55**(9):095105-1-095105-7. DOI: 10.1117/1.OE.55.9.095105

[19] Thornton DE, Spencer MF, Plimmer BT, Mao D. The digital holography demonstration: A table top setup for STEM-based outreach events. In: Proceedings Volume 10741, Optics Education and Outreach V; 107410J SPIE Optical Engineering + Applications; San Diego, California, United States: SPIE; 2018. DOI: 10.1117/12.2320380

[20] Salançon E, Escarguel A. Holography in education and popular science: A new versatile and vibrationless color device. European Journal of Physics. 2018;**40**(1):015301. DOI: 10.1088/1361-6404/aae8ba

[21] Xia H, Picart P, Montresor S, Guo R, Li JC, Yusuf Solieman O, et al. Mechanical behaviour of CAD/CAM occlusal ceramic reconstruction assessed by digital color holography. Dental Materials. 2018;**34**(8):1222-1234. DOI: 10.1016/j.dental.2018.05.007

[22] Casavola C, Lamberti L, Pappalettera G, Pappalettere C. Application of contouring to dental reconstruction. In: Jin H, Sciammarella C, Furlong C, Yoshida S, editors. Imaging Methods for Novel Materials and Challenging Applications, Volume 3. 2013. Conference Proceedings of the Society for Experimental Mechanics Series. New York, NY: Springer; DOI: 10.1007/978-1-4614-4235-6-25

[23] Song W, Huang K, Xi Y, Cho K. Interactive holography system for enhancing augmented reality experiences. Advanced Science Letters. 2015;**21**(3):354-357. DOI: 10.1166/asl.2015.5771

[24] Clini P, Quattrini R, Frontoni E, Pierdicca R, Nespeca R. Real/not real: Pseudo-holography and augmented reality applications for cultural heritage. In: Handbook of Research on Emerging Technologies for Digital Preservation and Information Modeling. IGI Global; 2017. DOI: 10.4018/978-1-5225-0680-5.ch009

[25] Berneth H, Burder F-K, Fäcke T, Hagen R, Hönel D, Rölle T, et al. Holographic recordings with high beam ratios on improved Bayfol HX photopolymer. In: Proceedings Volume 8776, Holography: Advances and Modern Trends III; 877603: SPIE Optics + Optoelectronics; Prague: Czech Republic. 2013. DOI: 10.1117/12.2018618

[26] Zanutta A, Orselli E, Fäcke T, Bianco A. Photopolymeric films with highly tunable refractive index modulation for high precision diffraction optics. Optical Materials Express. 2016;**6**(1):252-263. DOI: 10.1364/OME.6.000252

[27] Zhuk DI, Burunkova JA, Denisyuk IY, Miroshnichenko GP, Csarnovics I, Tóth D, et al. Peculiarities of photonic crystal recording in functional polymer nanocomposites by multibeam interference holography. Polymer. 2017; **112**:136-143. DOI: 10.1016/j.polymer.2017.02.004

[28] Pedrini G, Martinez-García V, Wiedmann P, Wenzelburger M, Killinger A, Weber U, et al. Residual stress analysis of ceramic coating by laser ablation and digital holography. Experimental Mechanics. 2016;**56**: 683-701. DOI: 10.1007/s11340-015-0120-3

[29] Ruiz CGT, De La Torre-Ibarra MH, Flores-Moreno JM, Frausto-Reyes C, Santoyo FM. Cortical bone quality

affectations and their strength impact analysis using holographic interferometry. Biomedical Optics Express. 2018;**9**(10):4818-4833. DOI: 10.1364/BOE.9.004818

[30] Kumar M, Shakher C. Experimental characterization of the hygroscopic properties of wood during convective drying using digital holographic interferometry. Applied Optics. 2016; **55**(5):960-968. DOI: 10.1364/AO.55.000960

[31] Kumar M, Birhman AS, Kannan S, Shakher C. Measurement of initial displacement of canine and molar in human maxilla under different canine retraction methods using digital holographic interferometry. Optical Engineering. 2018;**57**(9):094106-1-094106-09410612. DOI: 10.1117/1.OE.57.9.094196

[32] Wang D, Zhao Y, Rong L, Wan M, Shi X, Wang Y, et al. Expanding the field-of-view and profile measurement of covered objects in continuos-wave terahertz reflective digital holography. Optical Engineering. 2019;**58**(2): 023111-1-023111-7. DOI: 10.1117/1.OE.58.2.023111

[33] Montrésor S, Memmolo P, Bianco V, Ferraro P, Picart P. Comparative study of multi-look processing for phase map de-noising in digital Fresnel holographic interferometry. Journal of the Optical Society of America A. 2019;**36**(2):A59-A66. DOI: 10.1364/JOSAA.36.000A59

[34] Sun J, Chen Q, Zhang Y, Zuo C. Optimal principal component analysis-based numerical phase aberration compensation method for digital holography. Optics Letters. 2016; **41**(6):1293-1296. DOI: 10.1364/OL.41.001293

[35] Yaroslavsky L. Digital Holography and Digital Image Processing: Principles, Methods, Algorithms. New York: Springer; 2004. 584 p. ISBN: 978-1441953971

[36] Memmolo P, Bianco V, Paturzo M, Ferraro P. Numerical manipulation of digital holograms for 3-D imaging and display: An overview. Proceedings of the IEEE. 2017;**105**(5):892-905. DOI: 10.1109/JPROC.2016.2617892

[37] Shimobaba T, Ito T. Random phase-free computer-generated hologram. Optics Express. 2015;**23**(7):9549-9554. DOI: 10.1364/OE.23.00949

[38] Sang X, Gao X, Yu X, Xing S, Li Y, Wu Y. Interactive floating full-parallax digital three-dimensional light-field display based on wavefront recomposing. Optics Express. 2018; **26**(7):8883-8889. DOI: 10.1364/OE.26.008883

[39] Shimobaba T, Nakayama H, Masuda N, Ito T. Rapid calculation algorithm of Fresnel computer-generated-hologram using look-up table and wavefront-recording plane methods for three-dimensional display. Optics Express. 2010;**18**(19):19504-19509. DOI: 10.1364/OE.18.019504

[40] Shimobaba T, Sato Y, Miura J, Takenouchi M, Ito T. Real time digital holographic microscopy using the graphic processing unit. Optics Express. 2008;**16**(16):11776-11781. DOI: 10.1364/OE.16.011776

[41] Kimand S-C, Kim E-S. Effective generation of digital holograms of three dimensional objects using a novel look-up table method. Applied Optics. 2008; **47**(19):D55-D62. DOI: 10.1364/AO.47.000D55

[42] Kim S-C, Kim E-S. Fast computation of hologram patterns of a 3D object using run-length encoding and novel look-up table methods. Applied Optics. 2009;**48**(6):1030-1041. DOI: 10.1364/AO.48.001030

[43] Kang H, Yamaguchi T, Yoshikawa H, Kim S-C, Kim E-S. Acceleration method of computing a compensated phase-added stereogram on a graphic processing unit. Applied Opics. 2008; **47**(31):5784-5789. DOI: 10.1364/AO.47.005784

[44] Kang H, Yaras F, Onural L. Graphics processing unit accelerated computation of digital holograms. Applied Optics. 2009;**48**(34): H137-H143. DOI: 10.1364/AO.48.00H137

[45] Leseberg D, Fre´re C. Computer generated holograms of 3D objects composed of tilted planar segments. Applied Optics. 1988;**27**:3020-3024. DOI: 10.1364/AO.27.003020

[46] Tommasi T, Bianco B. Computer-generated holograms of tilted planes by a spatial frequency approach. Journal of the Optical Society of America A. 1993; **10**:299-305. DOI: 10.1364/JOSAA.10.000299

[47] Delen N, Hooker B. Free space beam propagation between arbitrarily oriented planes based on full diffraction theory: A fast Fourier transform approach. Journal of the Optical Society of America A. 1998;**15**:857-867. DOI: 10.1364/JOSAA.15.000857

[48] Esmer GB. Computation of holographic patterns between tilted planes [thesis]. Ankara: Bilkent University; 2004

[49] Matsushima K. Computer generated holograms for three-dimensional surface objects with shade and texture. Applied Optics. 2005;**44**(22): 4607-4614. DOI: 10.1364/AO.44.004607

[50] Yamamoto K, Senoh T, Oi R, Kurita T. 8k4k-size computer generated hologram for 3-D visual system using rendering technology. In: 4th International Universal Communication Symposium (IUCS), 18-19 October. IEEE; 2010

[51] Ahrenberg L, Benzie P, Magnor M, Watson J. Computer generated holograms from three dimensional meshes using an analytic light transport model. Applied Optics. 2008;**47**(10): 1567-1574. DOI: 10.1364/AO.47.001567

[52] Kim H, Hahn J, Lee B. Mathematical modelling of triangle-mesh-modelled three-dimensional surface objects for digital holography. Applied Optics. 2008;**47**(19):D117-D127. DOI: 10.1364/AO.47.00D117

[53] Liu Y-Z, Dong J-W, Chen B-C, He H-X, Wang H-Z. High-speed full analytical holographic computations for true-life scenes. Optics Express. 2010; **18**(4):3345-3351. DOI: 10.1364/OE.18.003345

[54] Lee W, Im D, Paek J, Hahn J, Kim H. Semi-analytic texturing algorithm for polygon computer-generated holograms. Optics Express. 2014;**22**(25): 31180-31191. DOI: 10.1364/OE.22.031180

[55] Su P, Cao W, Ma J, Cheng B, Liang X, Cao L, et al. Fast computer-generated hologram generation method for three-dimensional point cloud model. Journal of Display Technology. 2016;**12**(12): 1688-1694. DOI: 10.1109/JDT.2016.2553440

[56] Haist T, Schönleber M, Tiziani HJ. Computer-generated holograms from 3D-objects written on twisted-nematic liquid crystal displays. Optics Communications. 1997;**140**:299-308. DOI: 10.1016/S0030-4018(97)00192-2

[57] Yu L, Cai L. Iterative algorithm with a constraint condition for numerical reconstruction of a three-dimensional object from its hologram. Journal of the Optical Society of America A. 2001;

18(5):1033-1045. DOI: 10.1364/JOSAA.18.001033

[58] Rosen J, Brooker G. Digital spatially incoherent Fresnel holography. Optics Letters. 2007;**32**(8):912-914. DOI: 10.1364/OL.32.000912

[59] Muffoletto RP, Tyler JM, Tohline JE. Shifted Fresnel diffraction for computational holography. Optics Express. 2007;**15**(9):5631-5640. DOI: 10.1364/OE.15.005631

[60] Abookasis D, Rosen J. Computer-generated holograms of three-dimensional objects synthesize from their multiple angular viewpoints. Journal of the Optical Society of America A. 2003;**20**(8):1537-1545. DOI: 10.1364/JOSAA.20.001537

[61] Shi L, Huang F-C, Lopes W, Matusik W, Luebke D. Near-eye light field holographic rendering with spherical waves for wide field of view interactive 3D computer graphics. Journal ACM Transactions on Graphics. 2017;**36**(6):236-1-236-17. DOI: 10.1145/3130800.3130832

[62] Park J-H. Recent progress in computer-generated holography for three-dimensional scenes. Journal of Information Display. 2017;**18**(1):1-12. DOI: 10.1080/15980316.2016.1255672

[63] Chang C, Xia J, Jiang Y. Holographic image projection on tilted planes by phase-only computer generated hologram using fractional Fourier transform. Journal of Display Technology. 2014;**10**(2):107-113. DOI: 10.1109/JDT.2013.2285174

[64] Gao C, Liu J, Li X, Xue G, Jia J, Wang Y. Accurate compressed look up table method for CGH in 3D holographic display. Optics Express. 2015;**23**(26):33194-33204. DOI: 10.1364/OE.23.033194

[65] Murano K, Shimobaba T, Sugiyama A, Takada N, Kakue T, Oikawa M, et al. Fast computation of computer-generated hologram using Xeon Phi coprocessor. Computer Physics Communications. 2014;**185**(10):2742-2757. DOI: 10.1016/j.cpc.2014.06.010

[66] Jackin BJ, Miyata H, Baba T, Ohkawa T, Ootsu K, Yokota T, et al. A decomposition method for fast calculation of large scale CGH on distributed machines. In: Laser Applications to Chemical, Security and Environmental Analysis; 13-17 July 2014; Seattle, Washington, USA. 2014. DOI: 10.1364/AIO.2014.JTu4A.51

[67] Zhao Y, Cao L, Zhang H, Tan W, Wu S, Wang Z, et al. Time-division multiplexing holographic display using angular-spectrum layer-oriented method. Chinese Optics Letters. 2016;**14**(1):010005-1-010005-5. DOI: 10.3788/COL201614.010005

[68] Zhao Y, Cao L, Zhang H, Kong D, Jin G. Accurate calculation of computer-generated holograms using angular-spectrum layer-oriented method. Optics Express. 2015;**23**(20):25440-25449. DOI: 10.12364/OE.23.025440

[69] Kovachev M, Ilieva R, Benzie P, Esmer GB, Onural L, Watson J, et al. Holographic 3DTV displays using spatial light modulators. In: Onural L, Ozaktas HM, editors. Three-Dimensional Television: Capture, Transmission, Display. Heidelberg: Springer-Verlag Berlin; 2008. pp. 529-556. DOI: 978-3-540-72532-9

[70] Katkovnik V, Astola J, Egiazarian K. Discrete diffraction transform for propagation, reconstruction, and design of wavefield distributions. Applied Optics. 2008;**47**(19):3481-3493. DOI: 10.1364/AO.47.003481

[71] Katkovnik V, Migukin A, Astola J. Backward discrete wave field

propagation modelling as an inverse problem: Toward reconstruction of wave field distributions. Applied Optics. 2009;**48**(18):3407-3423. DOI: 10.1364/AO.48.003407

[72] Esmer GB. Fast computation of Fresnel diffraction field of a three dimensional object for a pixelated optical device. Applied Optics. 2013; **52**(1):A18-A25. DOI: 10.1364/AO.52.000A18

[73] Esmer GB. Performance assessment of a fast and accurate scalar optical diffraction computation algorithm. 3D Research. IEEE; 2013;**4**(1):1-7. DOI: 10.1007/3DRes.01(2013)2

[74] Esmer GB. Algorithms for fast calculation of scalar optical diffraction field on a pixelated display device. In: IEEE AFRICON 2013; 9-12 September 2013; Mauritius: IEEE; 2013. DOI: 10.1109/AFRCON.2013.6757704

[75] Esmer GB. Real-time computation of diffraction fields for pixelated spatial light modulators. Optics Express. 2015; **23**(10):12636-12647. DOI: 10.1364/OE.23.012636

[76] HoloEye. PLUTO phase only spatial light modulator reflective [Internet]. 2015. Available from: http://holoeye.com/spatial-light-modulators/slm-pluto-phase-only/

[77] Gander W, Gautschi W. Adaptive quadrature revisited. BIT Numerical Mathematics. 2000;**40**(1):84-101. DOI: 10.1023/A:1022318402393

Chapter 8

Fringe Pattern Analysis in Wavelet Domain

Yassine Tounsi, Abdulatef Ghlaifan, Manoj Kumar, Fernando Mendoza-Santoyo, Osamu Matoba and Abdelkrim Nassim

Abstract

We present a full-field technique for single fringe pattern analysis based on wavelet transform. Wavelets technique is a powerful method that quantifies at different scales how spatial energy is distributed. In the wavelets domain, fringe pattern analysis requires spatial modulation by a high-frequency carrier. We realize the modulation process numerically by combining the fringe pattern and its quadrature generated analytically by spiral phase transform. The first application concerns the speckle denoising by thresholding the two-dimensional stationary wavelet transform (2D-swt) coefficients of the detail sub-bands. In the second application, the phase derivatives are estimated from the 1D-continuous wavelet transform (1D-cwt) and 2D-cwt analysis of the modulated fringe pattern by extracting the extremum scales from the localized spatial frequencies. In the third application, the phase derivatives distribution is evaluated from the modulated fringe pattern by the maximum ridge of the 2D-cwt coefficients. The final application concerns the evaluation of the optical phase map using two-dimensional discrete wavelet transform (2D-dwt) decomposition of the modulated fringe pattern. The optical phase is computed as the arctangent function of the ratio between the detail components (high-frequency sub-bands) and the approximation components (low-frequency sub-bands). The performance of these methods is tested on numerical simulations and experimental fringes.

Keywords: continuous wavelets transform (cwt), stationary wavelets transform (swt), discrete wavelets transform (dwt), fringe pattern analysis, residual speckle denoising

1. Introduction

The single fringe pattern is an emerging technique for the analysis of full-field measurements in optical metrology. Fringe pattern analysis becomes a key technique in interferometric metrology which is more suitable for dynamic processes. The main purpose of this chapter is to exploit wavelet transform for single fringe pattern analysis to extract useful information. Wavelets are a powerful method allowing to know the spatial energy distribution at several scales.

Fringe pattern analysis using wavelets transform requires a spatial modulation by a high-frequency carrier as the case of Fourier domain [1]. We realize the modulation process digitally [2] by combining the fringe pattern and its quadrature generated analytically by spiral phase transform SPT [3, 4]. The advantage of this method lies in its implementation of numerical algorithms to reduce experimentation burden present in the phase shifting methods [5], namely, multiple fringe pattern generation and the experimental carrier introduction.

In the first application, we describe a residual speckle noise reduction technique by 2D-stationary wavelet decomposition of the speckle correlation fringe pattern [6]. The speckle noise is removed by thresholding the 2D-swt detail sub-bands coefficients [7].

The second application presents a method for phase derivative estimation based on 1D-continuous wavelet analysis of the modulated fringe pattern using Paul's function as the mother wavelet [8]. The phase derivative is calculated by extracting the extremum scales from the localized spatial frequencies. The third application is phase derivatives evaluation by 2D-continuous wavelet analysis of the modulated fringe pattern using complex Morlet's function as the mother wavelet [9]. The phase derivative is estimated by a maximum ridge of wavelets coefficients. Finally, we introduce an algorithm for optical phase extraction by 2D-Discrete wavelet decomposition of the modulated fringe pattern using Gabor's function as the mother wavelet [10]. The optical phase is extracted as the ratio between the detail components (high-frequency sub-bands) and the approximation components (low-frequency sub-bands).

The remaining part of the chapter proceeds as follow: divided into three sections: the second section will examine a brief description of continuous, discrete and stationary wavelet transforms. In the third section, we briefly introduce the speckle correlation fringes obtained by digital speckle pattern interferometry (DSPI). Finally, the fourth section is devoted to applications of wavelets to fringe pattern analysis.

2. Wavelet transform analysis

The concept of the wavelets and its numerous fields of the application make them a useful tool in several studies concerning localized variations analysis for non-stationary or transient signal analysis. This concept is based on the multiresolution analysis that represents signal variations at different scales. A detailed review of wavelets theory has been published by Ingrid Daubechies in [11]. However, we present here a brief description of three wavelets families for completeness.

2.1 Continuous wavelet transform (cwt)

The analysis of a given signal by continuous wavelets transform concerns to decompose it into several basic functions named wavelets. They are oscillatory functions with a finite duration and having zero average value, also, they are characterized by irregularity and the good localization These properties of the continuous wavelets make them a superior basis for signals analysis with discontinuities. The wavelets are constructed by translating and dilating them other wavelet functions resulting in a self-similar wavelet's families as follows.

$$\psi_{s,\xi}(x) = \psi\left(s^{-1}(x - \xi)\right) \tag{1}$$

where s and ξ are respectively the scale and translation parameters. The wavelet coefficients obtained by 1D-cwt decomposition of the signal $f(x)$ are given by:

$$w(s,\xi) = \int_{-\infty}^{+\infty} f(x).\psi^*_{s,\xi}(x).dx = s^{-1/2} \int_{-\infty}^{+\infty} f(x).\psi^*\left(s^{-1}(x-\xi)\right).dx \qquad (2)$$

where $\psi^*_{s,\xi}(x)$ represents the conjugate of the wavelet $\psi_{s,\xi}(x)$ defined for each shift ξ and each scale s. Wavelet coefficients are the output of the correlation product between a signal and the mother wavelet for different values of dilatation, and this is interpreted as a measure of the local similarity between them.

The 2D-continuous wavelet coefficients are obtained by the computation of the correlation product between the input image (signal 2D) and the mother wavelet. In the 2D case, a parameter of orientation angle is added, and the 2D-cwt wavelet coefficients of an input image $f(x,y)$ are defined as:

$$w(\xi,\eta,s,\theta) = s^{-2} \iint f(x,y)\cdot\psi^*\left(s^{-1}R_\theta(x-\xi,y-\eta)\right)dxdy \qquad (3)$$

where ξ and η are respectively the translation parameters along x- and y- directions, s and θ are respectively the scale vector and rotation angle; ψ denotes the 2D mother wavelet and R_θ represents the conventional 2×2 rotation matrix corresponding to θ given by

$$R_\theta = (x.\cos\theta + y.\sin\theta, y.\cos\theta - x.\sin\theta) \qquad (4)$$

2.2 Discrete wavelets transform (dwt)

The discrete wavelet transform denoted by dwt is a fast computing algorithm; it was introduced by Mallat [12] in 1989 for signal or image decomposition into two important components called approximation and details.

The 1D-dwt decomposition is based on two functions called scaling and wavelets. These functions, i.e., the scaling and the mother wavelet functions are related to the impulse response of $h[n]$ and $g[n]$, respectively, as illustrated in **Figure 1**. The approximation (low-frequency) and details (high-frequency) obtained respectively by scaling and wavelets functions are the down-sampled outputs of the first filters $h[n]$ and $g[n]$.

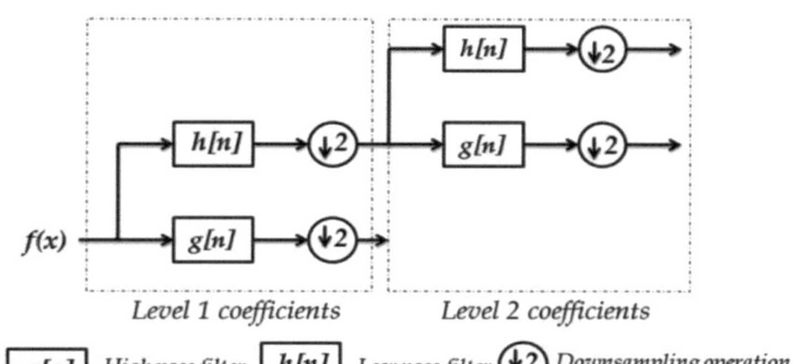

Figure 1.
Block diagram of filter analysis.

The relation between the two functions and the two filters is expressed as:

$$\phi(x) = \sqrt{2}\sum_n h[n]\varphi(2x - n) \tag{5}$$

$$\psi(x) = \sqrt{2}\sum_n g[n]\psi(2x - n) \tag{6}$$

g is the conjugate mirror of h:

$$g[n] = (-1)^n.h[1-n] \tag{7}$$

In Eq. (7), conjugation and mirror effects are represented respectively by $(-1)^n$and $(-n)$. The internal orthogonality relation is satisfied by the low-pass filter h as follows:

$$\sum_n h_n h_{n+2j} = 0 \tag{8}$$

and to have

$$\sum_n h_n^2 = 1 \tag{9}$$

The filter g satisfies h, the same internal orthogonality and both obey the mutual orthogonality relation expressed as following:

$$\sum_n h_n g_{n+2j} = 0 \tag{10}$$

In 2D-dwt, the wavelet transform of an image involves recursive filtering and sub-sampling process. **Figure 2** describes the procedure of dwt analysis, three detail images, and one approximation image are obtained at each level. Concerning detail images, they contain the high-frequency information we denote horizontal image sub-band by *HD*, vertical image sub-band by *VD* and the diagonal image sub-band by *DD* and the approximation image sub-band is denoted by *AI* which accommodates the low-frequency information.

The 2D scaling function denoted by $\phi(x,y)$, and the 3D wavelet $\psi^H(x,y)$, $\psi^V(x,y)$, and $\psi^D(x,y)$ are computed by the algebraic product between the one dimensional scaling and wavelet function as expressed in the following Eq. [13],

$$\begin{aligned} \phi(x,y) &= \phi(x).\phi(y) \\ \psi^H(x,y) &= \psi(x).\phi(y) \\ \psi^V(x,y) &= \phi(y).\psi(x) \\ \psi^D(x,y) &= \psi(x).\psi(y) \end{aligned} \tag{11}$$

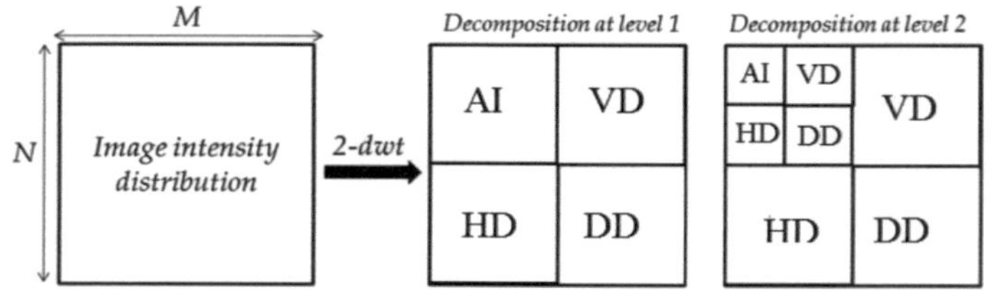

Figure 2.
Diagrams of dwt image decomposition.

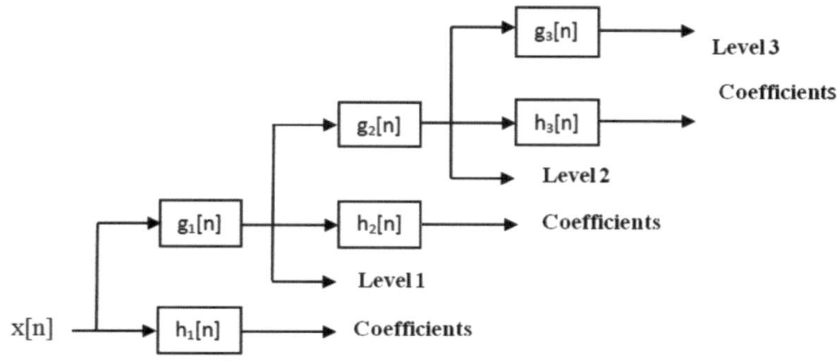

Figure 3.
Block diagram of filter analysis.

The three functions ψ^V, ψ^H, and ψ^D provide respectively the vertical, horizontal and diagonal variations, from there, the three details images are obtained.

2.3 Stationary wavelet transform (swt)

Stationary wavelets transform (swt) was presented in [14]. Swt has the same principle of decomposition as dwt transform, but the process of down-sampling is eliminated which means the swt is translation-invariant. The 2D-swt is founded on the idea of no down-sampling. Specifically, this transform is applied at each pixel of the image and saves the detail coefficients and exploits the low-frequency data at each level. A clear description of this technique is detailed in [15].

The principle of swt analysis is schematized in **Figure 3**. In brief, the swt technique consists to decompose an input sequence at each level and the given output is a new sequence that has the same length as the original sequences. In order to implement this transform, the process of original image decimation is removed, nonetheless, the size of the filter is modified at each level by zero paddings.

We introduce an operator denoted $\hat{Z}$ that alternates an input sequence with zeros so that for all the integers j, we have $(\hat{Z}_x)_{2j} = x_j$ and $(\hat{Z}_x)_{2j+1} = 0$. It is also assumed the filters H^r and G^r to have weights $\hat{Z}^r h$ and $\hat{Z}^r g$, respectively. Thus, the filter H^r is characterized by a weight $h^r{}_{2}{}^{r}{}_{j} = h_j$ and $h^r{}_{kj} = 0$ in the case where k is not a multiple of 2^r. The filter H^r is attained by introducing a zero between each adjacent pair of the filter $H^{(r-1)}$ elements, and an identical trend should be followed for G^r. The following relations satisfy the above statements:

$$D_0^r H^{[r]} = HD_0^r \text{ and } D_0^r G^{[r]} = GD_0^r \tag{12}$$

We start the swt definition by setting u^J to be the original input sequence, the stationary wavelet transform can be described for $j = J, J\text{-}1, \ldots, 1$,

$$u^{j-1} = H^{[J-j]} u^j \text{ and } v^{j-1} = G^{[J-j]} u^j \tag{13}$$

The vectors u^j and v^j acquire the same length for the vector u^J with a length of 2^J.

3. Speckle correlation fringe analysis

3.1 Speckle effect

The speckle appears as a granular structure and it results from the self-interference of a large number of coherent waves randomly scattered from and/or transmitted through a rough object surface [16]. When we illuminate a porous surface with coherent light, the scattered light intensity has a random spatial variation called "speckle effect." The "speckle pattern" appears chaotic and disordered; it is described by using statistics and probability. The speckle pattern structure is dependent on the coherence properties of the used illumination light and also on the characteristics of the object's surface diffusion. Locally in space, the electric field of the speckle pattern is given by computing the contribution sum of all illuminated scattering elements of the rough object's surface and is given by [16]

$$E(x,y,z) = \sum_{k=1}^{n} |a_k| \exp\left(j\phi_k\right) \quad (14)$$

In Eq. (14), n represents the scattering number of elements; a_k and ϕ_k are respectively the amplitude and phase of the kth scatterer contribution. The theorem of the central limit state that the random variable resulting from the sum of several independent random variables results in a Gaussian distribution when their number (number of an independent random variable) approaches infinity.

In the case where the number of scatterers is large, by applying the central Limit we deduce that:

a. The two components (real and imaginary) of the field are identically distributed Gaussian variables, independent and with zero means.

b. The intensity I(x,y,z) is characterized by a negative exponential probability distribution expressed as [16].

$$p(I) = \exp\left(-I/<I>\right)/<I> \quad (15)$$

The associated phase ϕ is uniformly distributed between $-\pi$ and π [16]

$$p(\phi) = 1/2\pi \quad (16)$$

3.2 Digital speckle pattern interferometry (DSPI)

Digital/electronic speckle pattern interferometry (DSPI/ESPI) is an optical interferometric technique demonstrated simultaneously by Macovski et al. [17], and Butters and Leendertz [18, 19] at the beginning of the 1970s. Later on, the technique was enriched by Beidermann and Ek [20] and Lokberg and Hogmoen [21] with several new investigations. **Figure 4** shows the typical schematic of the DSPI system with a sensitivity vector responding to out-of-plane deformation/displacement measurements (**Figure 4a**), and the digital electronics and image processing unit (**Figure 4b**). The DSPI technique measures the phase changes introduced by the speckle intensity changes. In this technique, a spatially filtered reference beam is added to the speckle pattern which is scattered from or transmitted through the test object, to code its phase. Therefore, a speckled interferogram, resulting from the superposition of the speckle pattern and the reference beams, contains essential information about the random phase. In DSPI, as shown in **Figure 4**, two speckle interferograms, corresponding to two different states of the object are recorded and stored in the frame grabber card. For example, one speckle interferogram, A, is recorded when the object is in its initial state, say the reference state, while another speckle interferogram, B, is recorded in the deformed state of the object, say by a small distance "*d*," as shown in **Figure 4a**. The DSPI fringe pattern is observed after subtraction of these two speckle interferograms A and B by using the appropriate

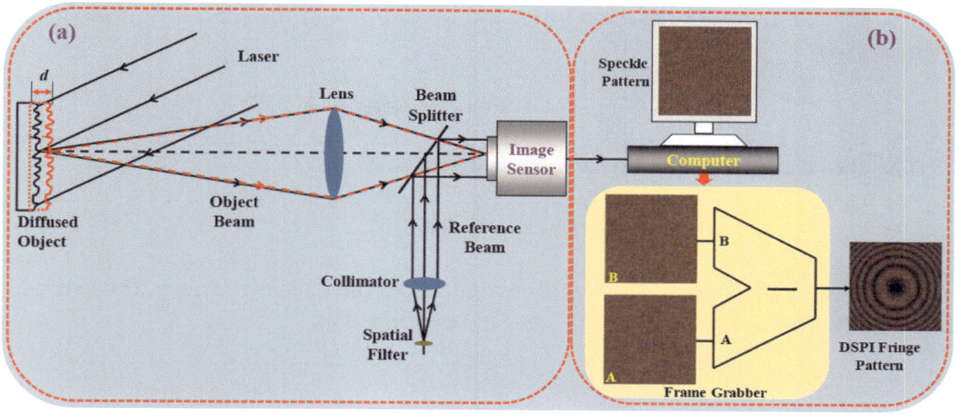

Figure 4.
Schematic of the DSPI system: (a) Speckle interferometer and (b) Digital electronics and image processing unit.

software. The useful information about the object is encoded in the DSPI fringe pattern like the one shown in **Figure 4b**.

The DSPI system can be made sensitive to out-of-plane or in-plane deformations, or both, depending on the design of the optical setup. Numerous measurement applications of DSPI have been investigated with different proposed experimental configurations [22–48] which have immense importance in the scientific, engineering, and industrial fields. The popularity of the technique, in optical metrology, is by virtue of the several advantages it offers such as: it is a non-contact/invasive type technique; it provides full-field of view information; it is faster in operation as almost real-time observations can be obtained; investigations on large-size objects can be observed with the use of proper optics, and the technique is less sensitive to environmental perturbations in comparison with its counterparts. The DSPI technique has proven its capability in many investigations, e.g., for the measurement of displacement/deformation with variable sensitivity to in-plane and out-of-plane direction [22–27], measurement of three-dimensional shape [28–30], surface roughness measurement [31], vibration measurement/monitoring [32–34], measurement of material properties [35–37], flow visualization [38, 39], measurement of refractive index and temperature distribution [40–42], and the investigation of the magnetic fields on the temperature profile of gaseous flames [43, 44].
A Doctoral Thesis describing the comprehensive study of DSPI and some of its novel investigations is recommended for interested readers working in this field [48].

Mathematically, the fringe formation in DSPI is demonstrated here. The intensity distribution recorded before the object deformation is expressed as follows:

$$f_B(x,y) = b(x,y) + a(x,y).\cos\phi_s(x,y) \tag{17}$$

In this equation, the term $b(x,y)$ represents the background or bias, $a(x,y)$ the visibility and term $\phi_s(x,y)$ is the original speckle phase appearing as a high frequency, and the intensity of pixel by pixel is randomly distributed. When the object is deformed, the intensity distribution of speckle-pattern becomes:

$$f_A(x,y) = b(x,y) + a(x,y).\cos(\phi_s + \varphi). \tag{18}$$

where $\varphi(x,y)$ represents the phase variation stemming from the object deformation. Assuming that the introduced deformations are sufficiently small, we can ignore the speckle decorrelation effects.

The speckle fringe correlation intensity distribution in the subtraction mode is expressed as:

$$f = f_A - f_B = 2.a.\sin(\varphi/2).\sin(\phi_s + \varphi/2) \tag{19}$$

The above equation can be further expressed as:

$$f^2 = 4a^2.\sin^2(\varphi/2).\sin^2(\phi_s + \varphi/2) \tag{20}$$

The speckle phase $\phi_s(x,y)$ changes faster across the speckle pattern, the second sine squared term has an ensemble average expressed as:

$$\sin^2(\phi_s + \varphi/2) \approx (1/2\pi n)\int_0^{2\pi} \sin^2(\phi_s + \varphi/2)\mathrm{d}(\phi_s + \varphi/2) = 1/2 \tag{21}$$

Hence, the speckle fringe correlation intensity distribution becomes:

$$f_C = f^2 = 2\mathrm{a}^2.\sin^2(\varphi/2) = \mathrm{a}^2.(1 - \cos\varphi) \tag{22}$$

The obtained intensity distribution f_c is rendered as the classical form of a cosine fringe pattern.

3.3 Fringe pattern analysis

The analysis of the fringe patterns concerns to evaluate the coded phase distribution related to the physical magnitude. We classified the phase extraction techniques to methods that explore a shifted fringe pattern as the phase shifting techniques and methods that require the introduction of a spatial carrier such as the Fourier transform method and the wavelet method.

The phase shifting (see Ref. [5]) and the Fourier transform methods (see Ref. [1]) are the most common methods exploited for fringe pattern analysis. The goal of the next section is to present algorithms based on wavelets transform to analyze a single frame speckle fringe pattern.

4. Applications to speckle correlation fringe analysis

As we stated previously, the fringe pattern analysis in wavelet domain requires a modulation process by introducing a spatial carrier. In our case, the spatial carrier in generated numerically and the entire process is carried out directly using a computer. Before presenting the four applications, we present in the next subsection the modulation process required by wavelets transform.

4.1 Fringe pattern modulation

In order to modulate a given fringes pattern in a chosen direction, we combine the fringe pattern and its quadrature with a $cos(m.x)$ and $sin(m.x)$ matrix respectively where m means the modulation rate. We consider the fringe pattern intensity distribution expressed as follow:

$$f(x,y) = b(x,y) + a(x,y)\cdot\cos(\varphi(x,y)) \tag{23}$$

A low-pass filter applied to the intensity distribution removes the background illumination and Eq. (23) becomes:

$$\tilde{f}(x,y) = a(x,y) \cdot \cos(\varphi(x,y)) \tag{24}$$

Larkin proposed recently a transform called spiral phase quadrature denoted - SPT- for the 2D fringe pattern (see Refs. [3, 4]). The SPT of $\tilde{f}$ is computed as:

$$SPT\left(\tilde{f}\right) = j \cdot \exp(j \cdot D(x,y)) \cdot a(x,y) \cdot \sin(\varphi) \tag{25}$$

where, $a(x,y).sin\varphi$ is the quadrature term, $j^2 = -1$, and $D(x,y)$ represents the direction map. Then, we obtain the sine fringe pattern as:

$$a(x,y) \cdot \sin(\varphi) = -j \cdot \exp(-j \cdot D(x,y)) \cdot SPT\left(\tilde{f}\right) \tag{26}$$

The direction map is computed by the ratio between the horizontal and vertical gradient of the $\tilde{f}$ as and analytically expressed as:

$$\tan(D) = \nabla_y \tilde{f} / \nabla_x \tilde{f} \tag{27}$$

So, we define the quadrature map as:

$$q(x,y) = b(x,y) \cdot \sin(\varphi) = -j \cdot \exp(-j \cdot D) \cdot SPT\left(\tilde{f}\right) \tag{28}$$

Then, the intensity distribution of the modulated fringe pattern is defined as:

$$f_m = \tilde{f} \cdot \cos(m \cdot x) - q \cdot \sin(m \cdot x) = b(x,y) \cdot \cos(\varphi + m \cdot x) \tag{29}$$

4.2 Speckle noise reduction using various wavelet-based techniques: a simulation study

4.2.1 Speckle noise reduction using 2D-swt

As a reminder, the speckle fringe pattern intensity distribution in the subtraction mode is expressed in (Eq. (19)) as $f = 2.a.\sin(\varphi/2).\sin(\phi_s + \varphi/2)$. The speckle fringes, presented in Eq. (19), are characterized by multiplicative residual speckle noise, therefore, a denoising step is necessary before the evaluation of the phase distribution and their derivatives. The high-frequency $sin(\phi_s + \varphi/2)$ noise should be removed with an appropriate filtering technique. After decomposition by swt of the speckle fringe pattern, the noise is reduced by thresholding the detail sub-bands coefficients using a soft threshold function. The choice of this type of thresholding stems from its characteristic to obtain near-optimal minimax rate. Generally, the threshold estimation requires the knowledge of the noise variance and the optimal threshold is estimated by taking into consideration the variance (σ^2) of the coefficients in the highest wavelets decomposition. **Figure 5** illustrates the capability of stationary wavelet transform thresholding technique to reduce the speckle noise. The left and the right halves present the intensity distribution of speckle fringe correlation before and after denoising. It is clearly shown in the line profile plot in **Figure 5b** along the line AB, the influence of swt to remove the frequency from the noised image.

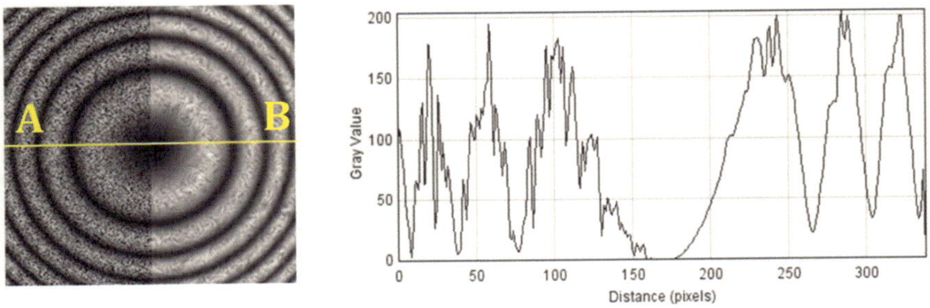

(a) (b)

Figure 5.
(a) Speckle fringe correlation before and after denoising; and (b) the 1D intensity profile along line AB.

4.2.2 Phase derivatives extraction using 1D-cwt

The idea is the phase extraction by spatial frequencies using the1D-continuous wavelet transform and Paul's function as the mother wavelet. The phase derivatives are evaluated from the modulus of the cwt coefficients by extracting the extremum scales from the localized spatial frequencies and the phase is obtained by numerical integration.

Generally, the modulated fringe pattern intensity distribution, as represented by Eq. (18), can also be expressed as:

$$f_m(x,y) = b(x,y) + a(x,y)\cos\left(m.x + \varphi(x,y)\right) \tag{30}$$

Using Eq. (2), the wavelet transform coefficients of the modulated fringe pattern, given in Eq. (30), can be represented as

$$w(x,s,\xi) = s^{-1/2}\int_{-\infty}^{+\infty}[b(x,y) + a(x,y)\cos\left(m.x + \varphi(x,y)\right)](\psi((y-\xi)/s))^{*}dy \tag{31}$$

The localization property of the wavelet is taken into account in order to write the Taylor series of the phase $\varphi(x,y)$ near the central value ξ as:

$$\varphi(x,y) = \varphi(x,\xi) + (y-\xi)\partial\varphi(x,y)/\partial y + \dots \tag{32}$$

Assuming that the bias a and the visibility b have a slow variation, so the higher order of $(y\text{-}\xi)$ is neglected with respect to the phase-modulated carrier because of the localization of the wavelet. With these considerations, the 1D-cwt coefficients are now computed as:

$$w(x,s,\xi) = s^{-1/2}b(x,\xi)a(x,\xi)\int_{-\infty}^{+\infty}\cos\left[mx + \varphi(x,\xi) + (y-\xi)\partial\varphi(x,\xi)/\partial y\right](\psi((y-\xi)/s))^{*}dy \tag{33}$$

and the Parseval's identity gives

$$w(x,s,\xi) = \frac{\sqrt{s}}{2\pi}\int_{-\infty}^{+\infty}\hat{f}_a(x,k)(\hat{\psi}(sk))^{*}e^{i\xi k}dk \tag{34}$$

where

$$\hat{f}_a(x,k) = b(x,\xi)a(x,\xi)\pi h(x,k) \tag{35}$$

with

$$\begin{aligned} h(x,k) = \delta(k-m_1)\exp\left[i(\varphi(x,\xi) - \xi\partial\varphi(x,\xi)/\partial y)\right] \\ +\delta(k+m_1)\exp\left[-i(\varphi(x,\xi) - \xi\partial\varphi(x,\xi)/\partial y)\right] \end{aligned} \tag{36}$$

$$m_1 = m + \partial\varphi(x,\xi)/\partial x \tag{37}$$

The wavelet transform reduces to.

$$\begin{aligned} w(x,s,\xi) = 0.5.a(x,\xi)b(x,\xi)\sqrt{s}[(\hat{\psi}(sm_1))^* \exp\left(i(\xi m + \varphi(x,\xi))\right) \\ +(\hat{\psi}(-sm_1))^* \exp\left(-i(\xi m + \varphi(x,\xi))\right)] \end{aligned} \tag{38}$$

Introducing Paul's mother wavelet defined as:

$$\psi(x) = \left(2^n n!(1-ix)^{-(n+1)}\right) / \left(2\pi\sqrt{(2n)!/2}\right) \tag{39}$$

we get

$$w(x,s,\xi) = ((2n)!)^{-1}a(x,\xi)b(x,\xi)s^{n+1/2}m_1^n \exp\left(-sm_1\right) \times \exp\left(i(\xi m + \varphi(x,\xi))\right) \tag{40}$$

whose modulus is computed as

$$|w(x,s,\xi)| = ((2n)!)^{-1}a(x,\xi)b(x,\xi)m_1^n s^{n+1/2} \exp\left(-sm_1\right) \tag{41}$$

The extremum scale denoted by S is represented by

$$S(x,\xi) = (2n+1)/2m_1 \tag{42}$$

Equations (37) and (42) provide the phase derivative as

$$\partial\varphi(x,\xi)/\partial y = (2n+1)/2S(x,\xi) - m \tag{43}$$

where m is the modulation ratio.

The performance of this algorithm is illustrated in **Figure 6**, where the phase derivatives of the speckle fringe correlation along x- and y-directions are presented.

4.2.3 Phase derivatives extraction using 2D-cwt

The idea is to evaluate the phase derivative by ridge point using2D-continuous wavelet transform and Morlet's function as the mother wavelet. The phase derivative is computed from maximums scales relating to the ridge point of the wavelet coefficient modules. From the modulated fringe pattern $f_m(x,y)$, we compute the 2D-cwt wavelet coefficients as defined in Eq. (3) and represented again as

$$w(\mathrm{t},\mathrm{d},s,\theta) = s^{-2}\iint f_m(x,y)\cdot\psi_M{}^*\left(s^{-1}R_\theta(x-t,y-d)\right)dxdy \tag{44}$$

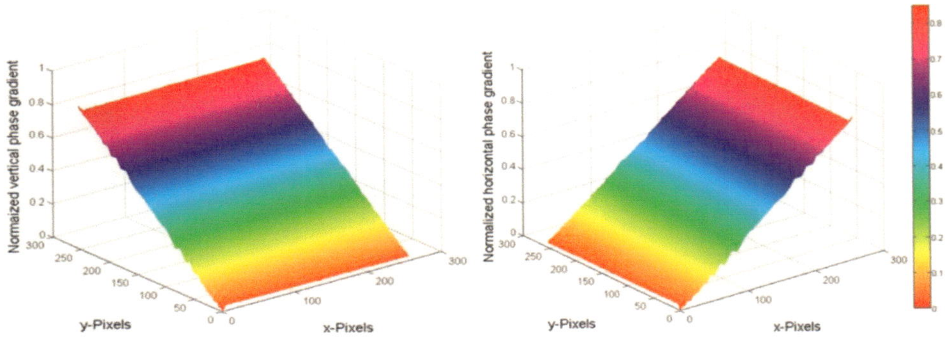

(a) (b)

Figure 6.
The evaluated phase derivatives along (a) x-direction *and (b)* y-direction *using 1D-cwt technique.*

The 2D complex Morlet is used as the mother wavelet: it is essentially a plane wave modulated by a Gaussian window, and is expressed as:

$$\psi_M = \exp\left(-\left(x^2 + y^2\right)/2\right) \cdot \exp\left(jk_0(x \cdot \cos\theta + y \cdot \sin\theta\right) \quad (45)$$

where k_O is the specific spatial frequency that should be in the range 5–6 in order to satisfy the admissibility condition [49], and $j^2 = -1$.

A new matrix containing the maximum value of each column of the wavelet coefficient modulus array defines the wavelet ridge, and then, the corresponding scale value is determined from the ridge wavelet [50, 51]. The maximum scales s_{max} correspond to the maximum ridge of the obtained wavelet coefficients modulus:

$$(s_{\max}, \theta) = \underset{s \in R^+,\, \theta \in [0,2\pi]}{\arg\max} |w(t, d, s, \theta)| \quad (46)$$

where s_{max} represents the scale value for maxima. The phase derivative is obtained by

$$\nabla\varphi = \left(k_0 + \left(k_0^2 + 2\right)^{1/2} / 2s_{\max}\right) - m \quad (47)$$

The horizontal and vertical phase derivatives evaluated from the speckle fringe correlation are presented in **Figure 7**.

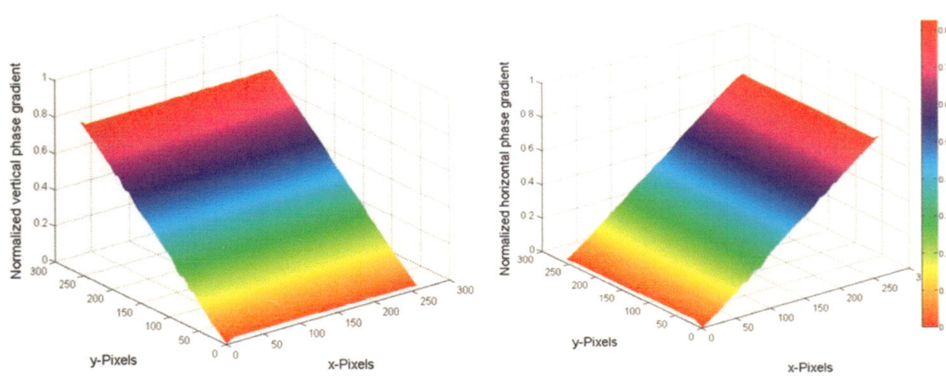

Figure 7.
The evaluated phase derivatives along (a) x-direction *and (b)* y-direction *using 2D-cwt technique.*

4.2.4 Optical phase extraction using 2D-dwt

This application concerns to extract the optical phase distribution by dwt components using the 2D discrete wavelets transform and Gabor's function as the mother wavelet. The 2D-dwt method decomposes the fringe pattern into two components: approximation (low-frequency) and details (high-frequency) using two principal quadrature mirror decomposition filters: low-pass and high-pass filters.

We consider the modulated fringes expressed in Eq. (30) as $f_m(x,y) = b(x,y) + a(x,y).\cos(m.x + \varphi(x,y))$. It depends on the desired phase $\varphi(x,y)$. **Figure 8** presents the four sub-bands of the modulated speckle fringe correlation after 2D-dwt decomposition.

The optical phase distribution coded in the modulated fringe pattern is evaluated by computing the arctangent function of the ratio between the detail and approximation images as following:

$$\varphi(x,y) = \text{atan2}((f_{\text{m}} * \text{h})/(f_{\text{m}} * \text{g})) \tag{48}$$

Inside the atan2 function, the numerator and denominator present the convolution product between f_m and the two filters h and g, and give the detail and approximation images respectively. An illustration of the two images is shown in **Figure 9**.

The detailed image in the numerator is chosen according to the direction of modulation. The low-pass filter h and high-pass filters g are respectively the real and imaginary part of Gabor function expressed as:

$$h = \exp\left[-\left(x^2/\sigma_x^2 + y^2/\sigma_y^2\right)/2\right].\cos\left(2\pi f_0(x\cos\theta + y\sin\theta)\right) \tag{49}$$

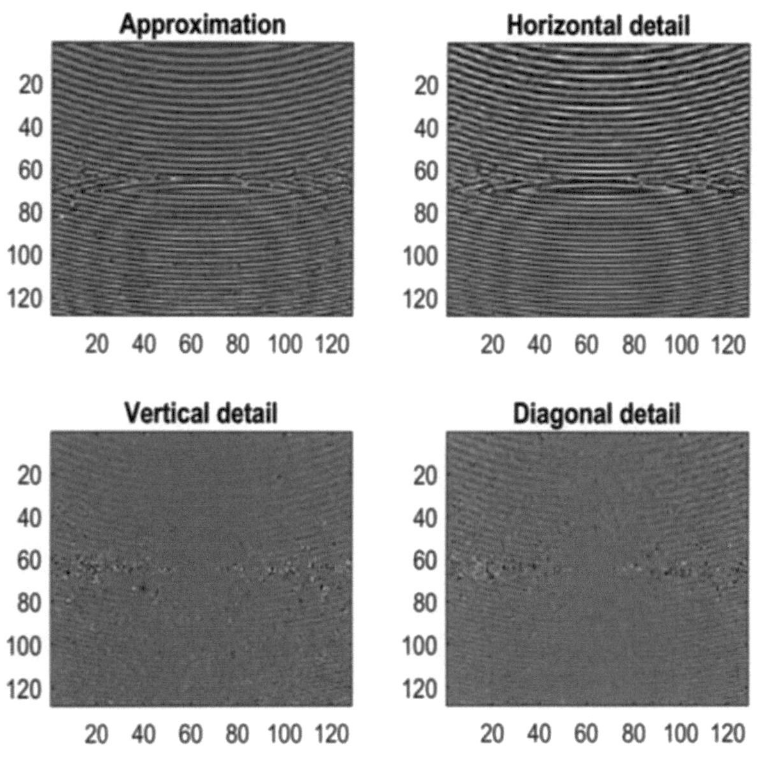

Figure 8.
The four sub-bands images after modulated fringe 2D-dwt decomposition.

(a) (b)

Figure 9.
Approximation and detail image of speckle fringe correlation.

$$g = \exp\left[-\left(x^2/\sigma_x^2 + y^2/\sigma_y^2\right)/2\right].\sin\left(2\pi f_0(x\cos\theta + y\sin\theta)\right) \tag{50}$$

The 2D Gabor's function [52] is the extension of the 1D Gabor function introduced by Gabor [53]. It is defined by a Gaussian window that modulates a complex exponential centered at a fixed frequency It is formulated by:

$$\mathrm{G} = \exp\left[-\left(\mathrm{x}^2/\sigma_\mathrm{x}^2 + \mathrm{y}^2/\sigma_\mathrm{y}^2\right)/2\right].\exp\left(j.2\pi f_0(x\cos\theta + y\sin\theta)\right) \tag{51}$$

Central frequency, orientation, spatial extent, and aspect are four essential parameters characterize Gabor function, and they can be adjusted by varying respectively f_0, θ, σ_x and σ_y. The use of the atan2 function in Eq. (48) provides the phase distribution modulo 2π as shown in **Figure 10a**. To remove this discontinuity, a phase unwrapping step is obligatory [54]; for this reason, we have implemented the PUMA algorithm [55]. The unwrapped phase illustration in three-dimensional representations is presented in **Figure 10b**.

4.3 Discussion of wavelet techniques on the numerically simulated fringe correlations

In order to study the performance and validity of the algorithms, an image quality assessment (Q) is used [56]. This metric model any image distortion by combining three factors. The first factor is the loss of correlation between the ideal

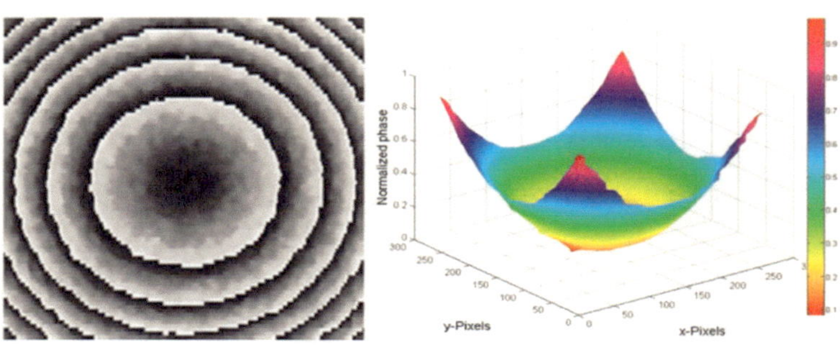

(a) (b)

Figure 10.
(a) Wrapped phase distribution retrieved by 2-dwt; and (b) continuous phase distribution volume visualization.

and the obtained images denoted O and E respectively, which measures the linear correlation degree between the two input images. This factor is defined as:

$$Q_1 = \sigma_{OE} / \sigma_O \sigma_E \tag{52}$$

The second factor is luminance distortion, it measures the mean luminance between O and E, which its value is computed as:

$$Q_2 = 2\overline{O}.\overline{E} / \left(\left(\overline{O}\right)^2 + \left(\overline{E}\right)^2 \right) \tag{53}$$

The third factor is contrast distortion that measures the contrast similarity between the two images and it is defined as:

$$Q_3 = 2\sigma_O \sigma_E / \left(\sigma_O^2 + \sigma_E^2\right) \tag{54}$$

And then, the Q coefficient is computed by the product of three factors:

$$Q = Q_1 \times Q_2 \times Q_3 \tag{55}$$

where $\overline{O}$and $\overline{E}$ present average of O and E, respectively, σ_O and σ_E are respectively the standard deviations of the two images. The Q index takes values between

Retrieved characteristic map	Q index value	Time in seconds
Horizontal phase gradient	0.95 (1-cwt) & 0.96 (2-cwt)	50.23 (1-cwt) & 20 (2-cwt)
Vertical phase gradient	0.95 (1-cwt) & 0.96 (2-cwt)	50.26 (1-cwt) & 20.11 (2-cwt)
Recovered phase distribution by 2-dwt	0.97	10.55

Table 1.
The measured Q index values.

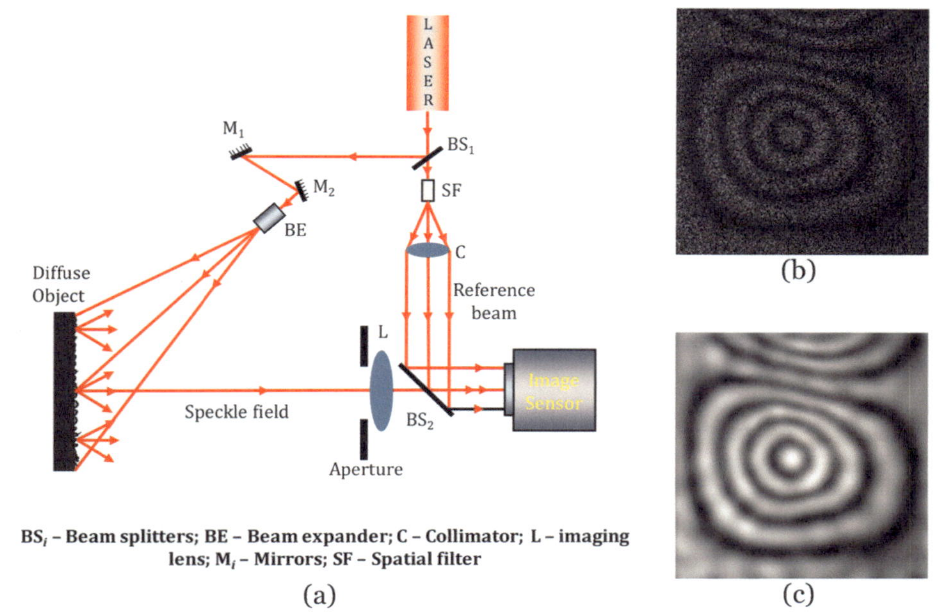

Figure 11.
(a) Schematic of DSPI setup [57, 58], (b) Recorded DSPI fringe pattern, and (c) Denoised DSPI fringe pattern by using swt thresholding technique.

−1 and 1 where 1 means that the retrieval characteristic is exact. The **Table 1** below summarizes the computed Q index values.

As a note, the 1D-cwt analyses images line by line, whereas the 2D-cwt scans images with the help of the parameter of orientation. On the analysis time side, we used Matlab on a 2.93 GHz Intel Pentium processor machine with 4 GB RAM. According to the Q index values, the presented algorithms give the desired information with good accuracy (Q between 0.95 and 0.97).

4.4 Application on an experimental speckle fringe correlation

After the presentation of the four applications of the stationary, continuous and discrete wavelet transform for fringe pattern analysis, and the favorable effectiveness validated by computer simulation, we exploit this algorithmic arsenal to analyze an experimental speckle fringe pattern recorded by using the DSPI setup showed in **Figure 11a**.

In DSPI, two specklegrams corresponding to the two different states (before and after deformation) of the object are recorded with an image sensor and then subtracted in order to get the speckle fringe correlation as shown in **Figure 11b**.

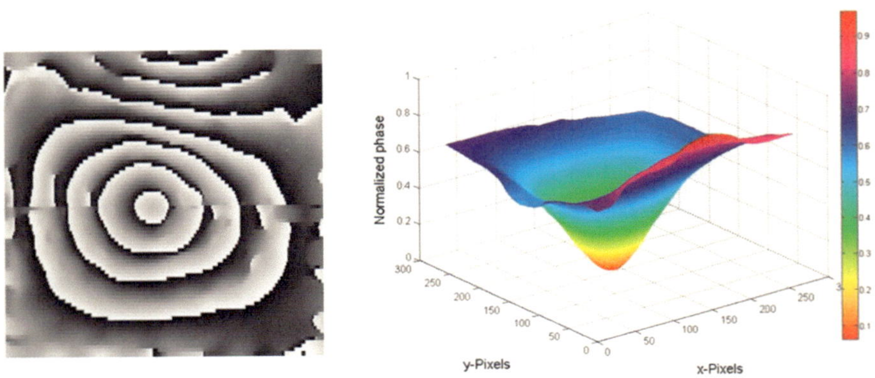

(a) (b)

Figure 12.
(a) Wrapped phase distribution extracted using dwt algorithm. (b) Three-dimensional representation of continuous phase distribution.

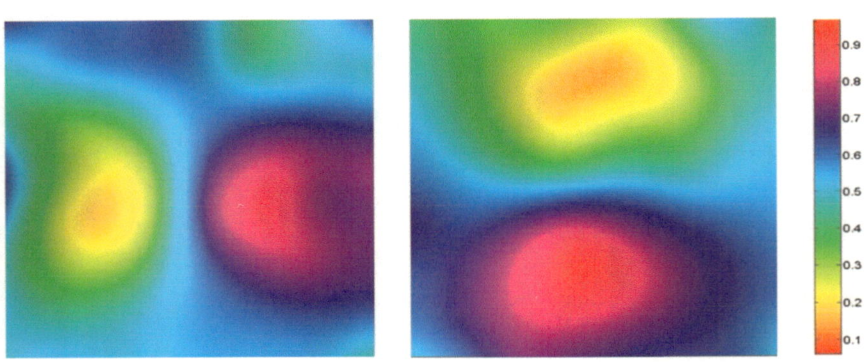

(a) (b)

Figure 13.
The estimated phase gradient distribution using 2-cwt technique along: (a) x-direction and (b) y-direction.

Figure 11c shows the denoised fringe resulting from the swt thresholding technique. The wrapped phase distribution and corresponding three-dimensional continuous phase distribution are shown in **Figures 12a** and **b**, respectively. The horizontal and vertical phase derivatives are shown in **Figure 13**.

5. Conclusion

This chapter has introduced an important thematic in interferometric metrology namely, fringe pattern analysis. We have discussed the capability of wavelets transform families (stationary, continuous and discrete wavelet) to analyze the fringe patterns.

This analysis consists in extracting the optical phase distribution coded in the fringes and its horizontal and vertical derivatives. We have presented with the help of computer simulations four applications of the wavelet transform, in order to analyze the speckle fringes correlation obtained in digital speckle pattern interferometry (DSPI). We summarize these applications as:

- Speckle noise reduction using stationary wavelet transform thresholding technique.
- Phase derivative extraction using one and two-dimensional continuous wavelet transform.
- Optical phase distribution extraction using the two-dimensional discrete wavelet transform

Acknowledgements

Manoj Kumar and Osamu Matoba acknowledge the support by Japan Society for the Promotion of Science KAKENHI (17F17370, 18H03888) and JST CREST Grant Number JPMJCR1755. All the authors wish to thank the anonymous reviewers for their great insight, comments, and suggestions.

Author details

Yassine Tounsi[1], Abdulatef Ghlaifan[1], Manoj Kumar[2*], Fernando Mendoza-Santoyo[3], Osamu Matoba[2] and Abdelkrim Nassim[1]

1 Department of Physics, Measurement and Control Instrumentation Laboratory IMC, Chouaib Doukkali University, El Jadida, Morocco

2 Department of Systems Science, Graduate School of System Informatics, Kobe University, Kobe, Japan

3 Centro de Investigaciones en Optica, Guanajuato, Mexico

*Address all correspondence to: manojklakra@gmail.com

References

[1] Takeda M, Ina H, Kobayashi S. Fourier-transform method of fringe-pattern analysis for computer-based topography and interferometry. Journal of the Optical Society of America. 1982; **72**(1, 1):156-160

[2] Barj EM, Afifi M, Idrissi AA, Rachafi S, Nassim K. A digital spatial carrier for wavelet phase extraction. Optik. 2005; **116**(11):507-510

[3] Larkin KG, Bone DJ, Oldfield MA. Natural demodulation of two-dimensional fringe patterns. I. General background of the spiral phase quadrature transform. Journal of the Optical Society of America A. 2001; **18**(8):1862-1870

[4] Larkin KG. Natural demodulation of two-dimensional fringe patterns. II. Stationary phase analysis of the spiral phase quadrature transform. Journal of the Optical Society of America A. 2001; **18**(8):1871-1881

[5] Creath K. Phase-shifting speckle interferometry. Applied Optics. 1985; **24**(18):3053-3058

[6] Barj EM, Afifi M, Idrissi AA, Nassim K, Rachafi S. Speckle correlation fringes denoising using stationary wavelet transform. Application in the wavelet phase evaluation technique. Optics and Laser Technology. 2006;**38**(7):506-511

[7] Donoho DL, Johnstone JM. Ideal spatial adaptation by wavelet shrinkage. Biometrika. 1994;**81**(3):425-455

[8] Afifi M, Fassi-Fihri A, Marjane M, Nassim K, Sidki M, Rachafi S. Paul wavelet-based algorithm for optical phase distribution evaluation. Optics Communication. 2002;**211**(1-6):47-51

[9] Ghlaifan A, Tounsi Y, Muhire D, Nassim A. Phase gradient retrieval from fringes pattern by using of two-dimensional continuous wavelet transforms. International Journal of Optics and Applications. 2017;7(4):69-74

[10] Ghlaifan A, Tounsi Y, Zada S, Muhire D, Nassim A. Two-dimensional discrete wavelets transform for optical phase extraction: Application on speckle correlation fringes. Optical Engineering. 2016;**55**(12):121708

[11] Daubechies I. Ten Lectures on Wavelets. Philadelphia, Pennsylvania: Society for Industrial and Applied Mathematics (SIAM); 1992. ISBN: 0-89871-274-2. https://books.google.co.jp/books?id=B3C5aG4OboIC&printsec=frontcover&dq=inauthor:%22Ingrid+Daubechies%22&hl=en&sa=X&ved=0ahUKEwixl ZLjnInjAhWFfd4KHX1hCCgQ6AEIKjAA#v=onepage&q&f=false

[12] Mallat SG. A theory for multiresolution signal decomposition: The wavelet representation. IEEE Transactions on Pattern Analysis and Machine Intelligence. 1989;(7):674-693

[13] Gonzalez RC, Woods RE. Digital Image Processing. 3rd ed. Upper Saddle River, N.J: Pearson; 2007

[14] Pesquet JC, Krim H, Carfantan H. Time-invariant orthonormal wavelet representations. IEEE Transactions on Signal Processing. 1996;**44**(8): 1964-1970

[15] Nason GP, Silverman BW. Lecture Notes in Statistics. Wavelets and Statistics. Vol. 103. New York: Springer; 1995. p. 281

[16] Goodman JW. Laser speckle and related phenomenon. In: Topics in Applied Physics. Vol. 9. Berlin: Springer; 1975

[17] Macovski A, Ramsey SD, Schaefer LF. Time-lapse interferometry and

contouring using television systems. Applied Optics. 1971;**10**(12):2722-2727

[18] Butters JN, Leendertz JA. Speckle pattern and holographic techniques in engineering metrology. Optics and Laser Technology. 1971;**3**:26-30

[19] Butters JN, Leendertz JA. Holographic and video techniques applied to engineering measurement. Measurement and Control. 1971;**4**(12): 349-354

[20] Biedermann K, Ek L. A recording and display system for hologram interferometry with low resolution imaging devices. Journal of Physics E: Scientific Instruments. 1975;**8**(7):571

[21] Løkberg OJ, Høgmoen K. Vibration phase mapping using electronic speckle pattern interferometry. Applied Optics. 1976;**15**(11):2701-2704

[22] Wykes C. Use of electronic speckle pattern interferometry (ESPI) in the measurement of static and dynamic surface displacements. Optical Engineering. 1982;**21**(3):213400

[23] Joenathan C, Franze B, Haible P, Tiziani HJ. Large in-plane displacement measurement in dual-beam speckle interferometry using temporal phase measurement. Journal of Modern Optics. 1998;**45**(9):1975-1984

[24] Fricke-Begemann T, Burke J. Speckle interferometry: Three-dimensional deformation field measurement with a single interferogram. Applied Optics. 2001; **40**(28):5011-5022

[25] Kumar M, Agarwal R, Bhutani R, Shakher C. Measurement of strain distribution in cortical bone around miniscrew implants used for orthodontic anchorage using digital speckle pattern interferometry. Optical Engineering. 2016;**55**(5):054101. DOI: 10.1117/1.OE.55.5.054101

[26] Kumar M, Agarwal R, Bhutani R, Shakher C. Deformation measurements in cortical bone-miniscrew interface in human maxilla by using digital speckle pattern interferometry. In: Speckle 2018, VII International Conference on Speckle Metrology; 7 September 2018; Vol. 10,834. International Society for Optics and Photonics; 2018. p. 1083412. DOI: 10.1117/12.2317714

[27] Pedrini G, Zou YL, Tiziani HJ. Simultaneous quantitative evaluation of in-plane and out-of-plane deformations by use of a multidirectional spatial carrier. Applied Optics. 1997;**36**(4): 786-792

[28] Creath K, Cheng YY, Wyant JC. Contouring aspheric surfaces using two-wavelength phase-shifting interferometry. Optica Acta: International Journal of Optics. 1985; **32**(12):1455-1464

[29] Moore AJ, Tyrer JR. Two-dimensional strain measurement with ESPI. Optics and Lasers in Engineering. 1996;**24**(5-6):381-402

[30] Yang L, Zhang P, Liu S, Samala PR, Su M, Yokota H. Measurement of strain distributions in mouse femora with 3D-digital speckle pattern interferometry. Optics and Lasers in Engineering. 2007; **45**(8):843-851

[31] Lehman MM, Pomarico JA, Torroba RD. Digital speckle pattern interferometry applied to a surface roughness study. Optical Engineering. 1995;**34**(4):1148-1153

[32] Creath K, Slettemoen GÅ. Vibration-observation techniques for digital speckle-pattern interferometry. Journal of the Optical Society of America A. 1985;**2**(10):1629-1636

[33] Wang WC, Hwang CH, Lin SY. Vibration measurement by the time-averaged electronic speckle pattern

interferometry methods. Applied Optics. 1996;**35**(22):4502-4509

[34] Chen F, Griffen CT, Allen TE. Digital speckle interferometry: Some developments and applications for vibration measurement in the automotive industry. Optical Engineering. 1998;**37**(5):1390-1398

[35] Ganesan AR. Measurement of poisson's ratio using real-time digital speckle pattern interferometry. Optics and Lasers in Engineering. 1989;**11**(4):265-269

[36] Kumar M, Gaur KK, Shakher C. Measurement of material constants (Young's modulus and Poisson's ratio) of polypropylene using digital speckle pattern interferometry (DSPI). Journal of the Japanese Society for Experimental Mechanics. 2015;**15**(Special_Issue):s87-s91. DOI: 10.11395/jjsem.15.s87

[37] Kumar M, Khan GS, Shakher C. Measurement of elastic and thermal properties of composite materials using digital speckle pattern interferometry. In: SPECKLE 2015, VI International Conference on Speckle Metrology; 24 August 2015; Vol. 9660. International Society for Optics and Photonics; 2015. p. 966011. DOI: 10.1117/12.2196390

[38] Song Y, Wu Y, Kulenovic R, Guo Z, Groll M. Digital speckle pattern interferometry (DSPI) using optical polarization phase shift for measurement of temperature field in thermal flow. In: Optical Technology in Fluid, Thermal, and Combustion Flow III; 21 November 1997; Vol. 3172. International Society for Optics and Photonics; 1997, pp. 442-453

[39] Song Y, Zhang W, Wu Y, Yao X. Digital speckle technique applied to flow visualization. Tsinghua Science and Technology. 2000;**5**(1):89-95

[40] Barbosa EA, dos Santos SC. Refractive and geometric lens characterization through multi-wavelength digital speckle pattern interferometry. Optics Communication. 2008;**281**(5):1022-1029

[41] Kumar M, Kumar V, Shakher C. Measurement of temperature and temperature distribution in diffusion flames using digital speckle pattern interferometry. In: Eleventh International Conference on Correlation Optics; 17 December 2013; Vol. 9066. International Society for Optics and Photonics; 2013. p. 90660Y. DOI: 10.1117/12.2047634

[42] Kumar M, Shakher C. Measurement of temperature and temperature distribution in gaseous flames by digital speckle pattern shearing interferometry using holographic optical element. Optics and Lasers in Engineering. 2015; **73**:33-39. DOI: 10.1016/j.optlaseng.2015.04.002

[43] Kumar M, Agarwal S, Kumar V, Khan GS, Shakher C. Experimental investigation on butane diffusion flames under the influence of magnetic field by using digital speckle pattern interferometry. Applied Optics. 2015; **54**(9):2450-2460. DOI: 10.1364/AO.54.002450

[44] Kumar M, Shakher C. Holographic optical element based digital speckle pattern shearing interferometer. In: Proceedings of SPIE—The International Society for Optical Engineering (SPECKLE); 2018. Vol. 10834. p. 1083414. DOI: 10.1117/12.2318342

[45] Kumar M, Birhman AS, Kannan S, Shakher C. Measurement of initial displacement of canine and molar in human maxilla under different canine retraction methods using digital holographic interferometry. Optical Engineering. 2018;**57**(9):094106. DOI: 10.1117/1.OE.57.9.094106

[46] Kumar M, Shakher C. Experimental characterization of the hygroscopic properties of wood during convective

drying using digital holographic interferometry. Applied Optics. 2016; **55**(5):960-968. DOI: 10.1364/ AO.55.000960

[47] Kumar M, Shakher C. Measurement of hygroscopic strain in deodar wood during convective drying using lensless Fourier transform digital holography. In: Optical Micro-and Nanometrology VI2016. Vol. 9890. International Society for Optics and Photonics. 2016. p. 989007. DOI: 10.1117/12.2227464

[48] Kumar M. Some novel applications of digital speckle pattern interferometry and digital holographic interferometry (doctoral dissertation). Delhi: Indian Institute of Technology; 2016. Available from: http://eprint.iitd.ac.in/handle/ 12345678/7179

[49] Torrence C, Compo GP. A practical guide to wavelet analysis. Bulletin of the American Meteorological Society. 1998; **79**(1):61-78

[50] Liu H, Cartwright AN, Basaran C. Moire interferogram phase extraction: A ridge detection algorithm for continuous wavelet transforms. Applied Optics. 2004;**43**(4):850-857

[51] Li S, Su X, Chen W. Wavelet ridge techniques in optical fringe pattern analysis. Journal of the Optical Society of America A. 2010;**27**(6):1245-1254

[52] Daugman JG. Uncertainty relation for resolution in space, spatial frequency, and orientation optimized by two-dimensional visual cortical filters. Journal of the Optical Society of America A. 1985;**2**(7):1160-1169

[53] Gabor D. Theory of communication. Electrical Engineers-Part I: General. Journal of the Institution of. 1947; **94**(73):58

[54] Pritt MD, Ghiglia DC. Two-Dimensional Phase Unwrapping: Theory, Algorithms, and Software. New York: Wiley; 1998. ISBN-10: 0471249351

[55] Bioucas-Dias JM, Valadao G. Phase unwrapping via graph cuts. IEEE Transactions on Image Processing. 2007;**16**(3):698-709

[56] Wang Z, Bovik AC. A universal image quality index. IEEE Signal Processing Letters. 2002;**9**(3):81-84

[57] Tounsi Y, Kumar M, Nassim A, Santoyo FM. Speckle noise reduction in digital speckle pattern interferometric fringes by nonlocal means and its related adaptive kernel-based methods. Applied Optics. 2018;**57**(27):7681-7690. DOI: 10.1364/AO.57.007681

[58] Zada S, Tounsi Y, Kumar M, Mendoza-Santoyo F, Nassim A. Contribution study of monogenic wavelets transform to reduce speckle noise in digital speckle pattern interferometry. Optical Engineering. 2019;**58**(3):034109. DOI: 10.1117/1. OE.58.3.034109